Metal–Organic Framework Derived Materials

Metal–Organic framework (MOF)-based materials are used for functional applications due to their large surface area, high porosity, tunable structure, controllable morphology, and surface functionality. This book explores MOF-derived materials focusing on their structural features, synthesis methods, unique properties, and versatile applications, including the underlying chemistry. It covers research developments in the field of design of novel carbon materials and metal-based materials from MOF and their composites, including specific applications.

Features:

- Provides a comprehensive idea regarding the fundamental chemistry of MOF-derived materials.
- Covers all the aspects of MOF-derived materials with a focus on derived carbon/metal species and composite materials.
- Includes application of MOF-derived materials specifically for environment and energy.
- Conveys a sense of relation between structure, property, and application of MOF-derived materials.
- Explores developments in different types of MOF-derived materials.

This book is aimed at senior undergraduate, graduate students, and researchers in materials science, chemical engineering, and chemistry.

Emerging Materials and Technologies

Series Editor: Boris I. Kharissov

The *Emerging Materials and Technologies* series is devoted to highlighting publications centered on emerging advanced materials and novel technologies. Attention is paid to those newly discovered or applied materials with potential to solve pressing societal problems and improve quality of life, corresponding to environmental protection, medicine, communications, energy, transportation, advanced manufacturing, and related areas.

The series takes into account that, under present strong demands for energy, material, and cost savings, as well as heavy contamination problems and worldwide pandemic conditions, the area of emerging materials and related scalable technologies is a highly interdisciplinary field, with the need for researchers, professionals, and academics across the spectrum of engineering and technological disciplines. The main objective of this book series is to attract more attention to these materials and technologies and invite conversation among the international R&D community.

2D Semiconductors for Environmental Remediation
Edited by Honey John, Nisha T Padmanabhan, Sona Stanly and Jith C Janardhanan

Materials from Natural Sources: Structure, Properties, and Applications
Edited by Ramesh Gardas, Neha Patni, and Amita Chaudhary

Dielectric Materials for Capacitive Energy Storage
Edited By Haibo Zhang and Hua Tan

Multifunctional Coordination Materials for Green Energy Technologies
Edited by Ghulam Yasin, Anuj Kumar, Sajjad Ali, Tuan Anh Nguyen, and Saira Ajmal

Advancements in Nanomaterials for Energy Conversion and Storage
Edited by Piyush Kumar Sonkar and Vellaichamy Ganesan

2D Materials-Based Sensors: Technology and Applications
Vinod Kumar Khanna

Metal–Organic Framework Derived Materials: Design Strategies and Applications
Gomathi Nageswaran, Varsha M V, Arun Kumar Rajasekaran and M Shashank Rao

For more information about this series, please visit: www.routledge.com/Emerging-Materials-and-Technologies/book-series/CRCEMT

Metal–Organic Framework Derived Materials

Design Strategies and Applications

Gomathi Nageswaran, Varsha M V, Arun Kumar Rajasekaran and M Shashank Rao

CRC Press
Taylor & Francis Group
Boca Raton London New York

CRC Press is an imprint of the
Taylor & Francis Group, an **informa** business

Designed cover image: Authors

First edition published 2025
by CRC Press
2385 NW Executive Center Drive, Suite 320, Boca Raton FL 33431

and by CRC Press
4 Park Square, Milton Park, Abingdon, Oxon, OX14 4RN

CRC Press is an imprint of Taylor & Francis Group, LLC

ISBN: 9781032485768 (hbk)
ISBN: 9781032558110 (pbk)
ISBN: 9781003432357 (ebk)

DOI: 10.1201/9781003432357

Typeset in Times
by Newgen Publishing UK

Contents

PART 1 Metal–Organic Framework Derived Materials: Fundamentals

PART 2   Applications of MOF-Derived Materials

Preface

Metal–organic frameworks (MOFs) have gained enormous amount of research interest since its discovery in the early 1990s. The emerging class of MOF-based materials have been used for various functional applications due to their large surface area, high porosity, tunable structure, controllable morphology, surface functionality, etc. In this class of materials, the MOF-derived porous carbon and metal holds a significant place due to its excellent properties inherited from the precursor MOF. The use of MOF as a precursor and/or sacrificial template yields porous carbon with unique characteristics such as high electrical conductivity, large surface area, and ordered structure which overcome the limitations associated with carbon obtained from other sources. Furthermore, highly dispersed metal clusters or metal oxides with high surface area can also be derived from MOF, making this research area significant in the materials science.

This book aims to explore the science of MOF-derived materials, focusing on their structural features, synthesis methods, unique properties, and versatile applications. This is a fundamental approach to understand the chemistry of MOF-derived materials covering all the aspects of this research area in a comprehensive manner. Even though there are a number of literatures on the chemistry of MOF, the research area of MOF-derived materials is relatively less documented. In this context, this book gives an overview of both the fundamental understanding and the recent research developments in this field which is necessary to design novel carbon materials and metal-based materials from MOF with outstanding properties specific to applications. Furthermore, this book can be used as a reference for academia, especially for graduate and postgraduate students, faculty members, and research scholars and for industry professionals working in the area of materials science, inorganic chemistry, nanoscience and technology, and chemical engineering.

The book has two parts: the first part describes the structure, properties, and synthesis, whereas the second part focuses on various applications of MOF-derived materials.

Chapter 1 discusses the introductory idea about the class of metal–organic framework materials. This chapter provides an overview of MOF chemistry with special focus on its structure, properties, and timeline of significant research progress in this area. The broad classification of MOF materials consisting of pristine MOF, MOF-based composites, and MOF derivatives is discussed in detail in this part. Furthermore, the different methods of functionalization of MOF by modifying the metal nodes and organic ligands are also included.

Chapter 2 presents an extensive idea of MOF-derived materials, its classification, and the different synthesis methods reported in the literature. The chapter covers different types of MOF-derived carbon such as pure porous carbon, metal/metal oxide-porous carbon, and other carbon composites. Different types of carbon and metallic materials have been derived from MOF with a special focus on their application. Furthermore, the different synthesis methods for MOF-derived materials such as carbonization, pyrolysis, annealing, and calcination will also be discussed.

Chapter 3 discusses the different structures obtained for MOF-derived carbon such as hollow/yolk shell structure, hierarchical structure, and one-dimensional hollow and two-dimensional nanoribbon structures. The understanding of structure–property relationship helps to fabricate and functionalize novel materials specific for various applications. The controllable morphology of derived carbon and metallic materials is an important aspect that needs to be discussed. The effect of carbonization temperature and time on the morphology and properties of derived materials will be explained in detail in this part.

Chapter 4 focuses on the adsorption and storage applications of MOF-derived materials. In this chapter, the need for adsorption and storage of CO_2 and H_2 will be discussed. Various materials used for this application and their limitations will be discussed. The mechanism of adsorption and the recent developments in MOF-derived materials for adsorption and storage of CO_2 and H_2 will be explained in this chapter.

Chapter 5 elaborates the use of MOF-derived materials for various catalytic reactions, including electrocatalytic, photocatalytic, heterogeneous catalytic, and single atom catalytic reactions, and the recent developments of the same will be discussed.

Chapter 6 briefly surveys the importance of sensing biologically important analytes and various other analytes like metal ions, volatile organic carbon, NH_3, etc., causing environmental pollution and health hazards using MOF-derived materials as a sensing platform. The different sensing mechanisms, including electrochemical and colourimetric, will be discussed with more emphasis on electrochemical sensors. This chapter will provide an overview of the challenges and opportunities of applying various MOF-derived materials and their hybrids/composites as sensors of different mechanism.

Chapter 7 reviews the different forms of MOF-derived materials used for energy storage applications. This includes porous carbon, metal oxide, and their hybrid/composite and carbon matrix metal that are applied in rechargeable batteries, dye-sensitized solar cells, and supercapacitors. The challenges of MOF-derived material in energy applications will also be discussed in this chapter.

Chapter 8 focuses on the environmental applications of MOF-derived materials by discussing various environmental issues and technologies available to identify the pollutants and the cleaning mechanisms. The role of MOF-derived materials in sensing the various pollutants which have impact on environment and health of living beings, with a focus on sensing and removal mechanisms, will be explained in this chapter. Furthermore, the removal of these environmental pollutants through adsorption and catalysis will be discussed here.

Chapter 9 gives an idea of the emerging applications of MOF-derived materials in biomedical research. The challenges in the design of MOF-derived nanozymes and their biomedical applications, including imaging, antibacterial agents, cancer therapy, and biosensors, will be discussed here. MOF-templated polymers with hierarchically structured assemblies are another class of emerging materials that we will discuss in this chapter.

Author Biographies

Gomathi Nageswaran is Professor at the Indian Institute of Space Science and Technology, an autonomous institute under the Department of Space, Government of India. She received her PhD from the Indian Institute of Technology, Kharagpur, India, in 2009. Her current research interests include electrochemical sensing, carbon dioxide adsorption and catalytic applications of metal–organic framework-based materials, inorganic quantum dots for fluorescence sensors, membrane-based separation for water purification application, plasma surface modification of nanomaterials for enhanced electrochemical activity, and membrane performance.

Varsha M V received her MSc degree in Chemistry from Cochin University of Science and Technology, Kerala, India, in 2015. Currently, she is pursuing her PhD degree in Chemistry under the guidance of Dr Gomathi Nageswaran at the Indian Institute of Space Science and Technology (IIST), Thiruvananthapuram, India. Her research work includes development of novel materials based on metal–organic framework for the electrochemical sensing of analytes of environmental significance and development of electrocatalysts for water splitting reaction.

Arun Kumar Rajasekaran received his MPhil in Scientific Computing from the University of Cambridge and his bachelor's degree in Material Science from National Institute of Technology-Trichy, India. Currently, he is pursuing his PhD in Chemistry under the guidance of Prof. David Wales, FRS, and Dr Tom Bennett at the University of Cambridge, UK. His research focuses on the computational investigation of MOFs, emphasizing their carbon capture ability. He has spent his research stays in India, France, and China working on porous materials. In addition, Arun holds several prestigious research awards, including an Honorary Cambridge Nehru award (2020), a Hughes Hall scholarship (2021), Lundgren research award (2021), Education future award (2019), and several more. His research interests also include thermoelectric materials, polymer composites, network science, ancient mythology, and climate action.

M Shashank Rao received his BE degree in Mechanical Engineering from Chaitanya Bharathi Institute of Technology (Osmania University), Hyderabad, India and his MTech degree in Material Science and Technology from Indian Institute of Space Science and Technology, India. His research interests include the development of MOF- and BMOF-based adsorbents for CO_2 capture and sequestration. His other interests are design of novel catalysts for electrochemical reduction of CO_2.

Part 1

Metal–Organic Framework Derived Materials

Fundamentals

1 Metal–Organic Framework

A Brief Introduction

1.1 INTRODUCTION

In the late 1990s, the introduction of a novel type of porous compound with an inorganic-organic hybrid framework, known as metal-organic frameworks (MOFs), revolutionized the field of material science and gave rise to a new family of porous materials. MOFs are a class of porous coordination polymers (PCP), consisting of metal ions or clusters coordinated to organic ligands. Over the last two decades, MOFs have emerged as a highly interdisciplinary field, rooted in coordination and solid state chemistry. MOFs are created through the coordination of metal-containing units or secondary-building units (SBUs) with organic linkers, resulting in open frameworks that possess remarkable characteristics such as permanent porosity, stability, high surface area, and large pore volume. The porosity is achieved by employing lengthy organic linkers that provide significant storage space and a multitude of adsorption sites within the MOFs. Additionally, the pore structure of MOFs can be modified and tailored in a systematic manner. MOFs are characterized by their highly porous crystalline structure and have been extensively investigated for their tunable pore size, high surface area, and diverse range of potential applications, including gas storage and separation, catalysis, drug delivery, and sensing, among others. As a result, MOFs have emerged as an active area of research and development, continuing to draw significant attention in the scientific community. The better understanding of the structure of MOF and its underlying interactions is essential to design novel MOFs with characteristics suitable for specific applications. This chapter provides an overview of MOF chemistry with a special focus on its structure, properties, and timeline of significant research progress in this area. The broad classification of MOF materials consisting of pristine MOF, MOF-based composites, and MOF derivatives is discussed in detail in this part. Furthermore, the different methods of functionalization of MOF by modifying the metal nodes and organic ligands are also included.

1.2 BACKGROUND

According to the International Union of Pure and Applied Chemistry (IUPAC) definition, coordination polymers are coordination compounds with repeating coordination entities extending in one, two, or three dimensions, whereas coordination network

DOI: 10.1201/9781003432357-2

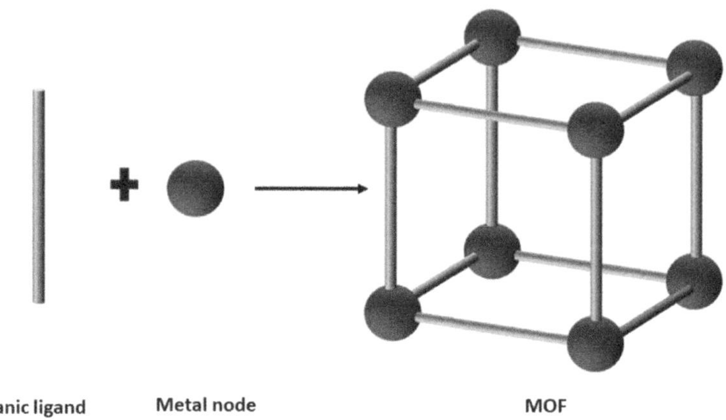

FIGURE 1.1   Coordination of metal- and ligand-forming MOF structure.

solids are a subcategory of coordination polymers, where a coordination compound extends through repeating its coordination entities in two or three dimensions. While "coordination polymer" is the more commonly used term, the IUPAC recognizes "coordination network" as a valid term as well. However, it is important to note that these two terms are not identical and that a coordination network is a specific type of coordination polymer in which the metal ions are linked together by bridging ligands to form an extended, often infinite, three-dimensional network. In contrast, a coordination polymer can refer to any polymer in which metal ions or clusters are linked together by organic ligands, regardless of the geometry or dimensionality of the resulting structure. Even though the research on porous coordination polymers started in 1964, the concept of MOF was first proposed in the late 1990s and the term MOF was coined by O. M. Yaghi [1]. An interpenetrating 3D network of $Cu(4,4'\text{-bpy})_{1.5}.NO_3(H_2O)_{1.25}$ cationic framework synthesized by a hydrothermal route was analysed using single crystal X-ray crystallography (SCXRD). The trigonal planar geometry of Cu(I) centre is a unique characteristic of the framework. Since then, the number of reports on MOFs has been increasing with a focus on design strategies, functionalization, and tuning its properties to make it a suitable candidate for different applications. In order to design or modify novel or existing MOF, basic understanding about the structure and underlying chemistry is essential (Figure 1.1).

1.3   SECONDARY BUILDING UNITS (SBUs)

The structure of MOF consists of two components: metal ions/cluster and organic ligands. SBUs are polynuclear inorganic clusters formed by the assembly of metal ion clusters to other metals through non-metal bond (i.e., M-O-M, M-O-C-O-M bonds, etc.) that give rise to periodic networks or structures. SBU is an important factor determining the topology, directionality, and stability of the framework. The substitution of metal in the SBU with a different metal atom leads to the formation of different MOF. Different SBU geometries are observed in MOF structures with different points of extension. For example, an octahedron SBU has six points of extension, and it can be found in MOFs such as MIL-100 and ZIF-8. A trigonal prism SBU also has six points of extension and can be found in MOFs such as MIL-101 and HKUST-1. A square paddlewheel SBU has four points of extension and can be found in MOFs such as MOF-74 and IRMOF-3. A triangle SBU has three points of extension and can be found in MOFs such as MOF-5 and UiO-66. Furthermore, the different geometries of SBUs can affect the properties such as porosity and stability of the resulting MOFs.

The choice of precursors is an important factor determining the stability of the framework, i.e., the selection of metal and ligand precursors according to HSAB (hard and soft acid and base) principle could lead to chemically stable MOFs, i.e., hard Lewis acid (Zr^{4+}, Ti^{4+}, Cr^{3+}) could form stable structures with hard Lewis base (oxygen-containing ligands such as carboxylate), whereas soft acid (Cu^{2+}, Ni^{2+}, Co^{2+}) could form stable structures with soft base (nitrogen-containing ligands). In short, SBUs are designed to provide two types of stability in MOF: thermodynamic and mechanical stability. The former is achieved by strong coordinate bond, whereas the latter is obtained from strong directional bond that can keep the position of metal sites intact in the framework. Therefore, in comparison with primary building units having mononuclear metal centres, SBUs provide more stable 3D framework structures. The coordination geometries based on mononuclear metal sites may often give rise to fragile structures which are prone to the reaction conditions.

1.4 COORDINATIVELY UNSATURATED SITES/OPEN METAL SITES

MOFs contain solvent molecules bound to the metal sites within the MOF pores. However, these solvent molecules can be removed through different methods either by heating in air or under vacuum. This removal process leaves behind open metal sites in the framework, resulting in more number of accessible reactive metal sites in MOF which can function as interaction sites with the guest molecules. $Cu_2(O_2CR_4)$ SBU with paddlewheel geometry is one of the simplest SBU studied that contains axially coordinated solvent molecules. In this, open metal sites are created by removing coordinated solvent species and has been used for the reversible binding of various gases such as H_2, CH_4, and C_2H_2. In a typical MOF, the metal ion which is part of the SBU only generates open metal sites and not the metal species belonging to the metallo-ligand (which is present in certain MOF). However, it is important to ensure that any changes made to the metal ions in the SBUs do not compromise the integrity of the MOF structure. This means that any ligand exchange reactions or other modifications made to the metal ions must be carefully designed to avoid collapse of the MOF structure and to preserve its crystallinity and porosity. Removing a labile terminal ligand from a metal ion to create an open metal site is one such modification. The discovery of MOFs with open metal sites was first reported in 1999. HKUST-1 was the first MOF to have an easily accessible and well-defined open metal site (OMS). Later in 2000, a Zn-based MOF termed as MOF-4 with OMS was reported. The OMSs in MOFs often represent the strongest binding sites within the framework because they are coordinatively unsaturated and can form strong interactions with guest molecules or other chemical species. This increased interaction with sorbate molecules can lead to enhanced selectivity and capacity for gas adsorption, separation, and catalysis compared to MOFs where the metal sites are fully occupied by ligands. In addition, the properties of OMS can be tuned by changing the nature of the metal ion, the type of ligand, and the degree of activation or functionalization of the MOF. For example, changing the metal ion from copper to zinc can affect the acidity and redox properties of the OMS, which in turn can affect the catalytic activity of the MOF.

1.5 METAL-LIGAND COORDINATION

The organic ligands are referred to as 'struts' or linkers and metal nodes as 'joints' in MOF structures. The appropriate choice of ligand and connection geometry of metal node gives the desired topology. Some examples of the different types of ligands used in MOF are: (i) carboxylate ligands; (ii) imidazolate-based ligands; (iii) pyridine-based ligands; and (iv) phosphonate ligands. These ligands can be combined in various ways to create MOFs with different structures and properties. Discrete di-, tri-, tetra-nuclear metal clusters with multidentate carboxylate linkers serve as SBU, provided, certain structural elements important for their polymerization into coordination network are present: (i) geometry of SBU is determined by those atoms which act as the point of extension to other SBU and further decides the topology of the network; (ii) substitution of mono-carboxylate with di-, tri-, or tetra-carboxylate ligands result in the extension of framework. Additionally, the orientation of ligands can impact the accessibility of the metal centre to other reactants or ligands, which can affect the reactivity of the material. Therefore, a basic understanding of the coordination mode of each ligand is critical in predicting the topology of the coordination network or the overall arrangement of the ligands and metal centres in the framework. Typically, coordination mode of a ligand refers to the specific way in which the ligand binds to the central metal ion, including the number of coordinating atoms, the geometry of the coordination sphere, and the orientation of the ligand relative to the metal centre; and (iii) the presence of open metal sites in the framework by removing the terminal ligands in the framework. Post-functionalization of OMSs in metal-organic frameworks (MOFs) represents a powerful approach for the introduction of new functionalities, such as ligand substitution via OMS coordination, which can effectively modify the properties of MOFs, including their catalytic activity, selectivity, and stability (Table 1.1).

1.6 CLASSIFICATION OF MOFs

MOFs can be classified based on their topologies, composition, and properties, and these classifications can be useful for understanding the structure-property relationships of MOFs. One way to classify MOFs is based on their topologies, which describe the connectivity and arrangement of metal ions and organic ligands in the framework, which is independent of the specific metal ions and ligands used in the synthesis. MOFs with the same topology have identical connectivity patterns but can have different metal ions and ligands, resulting in different properties.

Classification of MOFs based on composition can be done in different ways: based on the nature of metal or ligand precursors or using metal to ligand ratio. MOFs can be

TABLE 1.1
Examples of benchmark MOFs and their applications

MOF	Formula	Metal node	Ligand	Application	Ref.
MOF-5/ IRMOF-1	$Zn_4O(BDC)_3$	Zn^{2+}	1,4-Benzodicarboxylic acid	H_2 storage	[2]
MOF-16/ IRMOF-16	$Zn_4O(TPDC)_3 \cdot 17DEF.2H_2O$	Zn^{2+}	p-Terphenyl-4,4′-dicarboxylic acid	Drug delivery	[3]
MOF-199/ HKUST-1	$Cu_3(BTC)_2$	Cu^{3+}	1,3,5-Benzene tricarboxylic acid	Adsorption, separation, catalysis	[4]
UiO-66	$Zr_6O_6(BDC)_6$	Zr^{4+}	BDC	Catalysis, adsorption, supercapacitor	[5]
UiO-68	$Zr_6O_6(TPDC)_6$	Zr^{4+}	p-Terphenyl-4,4′-dicarboxylic acid	Catalysis, drug delivery, sensor	[6]
ZIF-8	$Zn(MIM)_2$	Zn^{2+}	2-Methyl imidazole	CO_2 capture, gas separation	[7]
ZIF-90	$Zn(FIM)_2$	Zn^{2+}	Imidazolate-2-carboxyaldehyde	Adsorption	[8]
MIL-53(Cr)		Cr^{3+}	BDC	Adsorption, separation, sensor	[9]
MIL-53(Al)	Al(OH)(BDC)	Al^{3+}	BDC	Adsorption	[10]

synthesized using transition metals, lanthanides, and main group metals. Transition metal-based MOFs are among the most widely studied and utilized due to their versatility and abundant coordination chemistry. Lanthanide-based MOFs have also gained attention in recent years due to their unique optical and magnetic properties. Main group metal-based MOFs, on the other hand, have been less explored but hold potential for applications in gas separation and catalysis.

Another way to classify MOFs based on composition is by their organic ligands. MOFs can be synthesized using a wide range of ligands, including carboxylic acids, amines, pyridines, and imidazoles. Carboxylic acid-based ligands are among the most widely used and studied due to their strong coordination ability and tunable properties. Amines, pyridines, and imidazoles are also commonly used as ligands due to their nitrogen donor atoms, which can form strong coordination bonds with metal ions.

MOFs can also be classified based on their metal-to-ligand ratio (MLR), which refers to the number of metal ions per ligand in the framework. MOFs with low MLR typically have open structures and high porosity, while MOFs with high MLR typically have dense structures and low porosity. Furthermore, MOFs can be classified based on the coordination geometry of the metal ions, which can affect their properties and applications. For example, MOFs with octahedral coordination geometry are often used as catalysts due to their ability to accommodate guest molecules in their inner cavities.

TABLE 1.2
MOF-based composites and their applications

MOF	Composite constituent	Application	Ref.
HKUST-1	Carbon	Electrochemical sensing	[11]
ZIF-8	PdO@ZnOSnO$_2$	Gas sensing	[12]
UiO-66-NH$_2$	ssDNA	Luminescent sensing	[13]
Cu-TCPP(Fe)	Glucose oxidase	Antimicrobial agent	[14]
Cu-BTC	CuS QDs	Electrocatalysis	[15]
MOF-525	Graphene oxide	Adsorption	[16]
2D Ni MOF	Graphene	Energy storage	[17]
Cu-BTC	Poly(2,6-dimethyl-1,4-phenylene oxide) (PPO)	Gas separation	[18]
MIL-53(Cr)	TMPTA	Dye removal	[19]
HKUST-1	Fe$_3$O$_4$	Drug delivery	[20]

Another important classification of MOF based on composition and structure consists of pristine MOF, MOF composites, and MOF derivatives. Pristine MOFs refer to MOFs in their original, unmodified state. These MOFs are typically synthesized by self-assembly of metal ions and organic ligands under specific reaction conditions which possess a well-defined structure and properties. MOF composites refer to materials that combine MOFs with other materials to create new materials with enhanced properties. MOF composites can be prepared by incorporating MOFs into a host matrix or by functionalizing MOFs with other materials. For example, MOFs can be incorporated into polymers, nanoparticles, or graphene to create hybrid materials with improved properties such as increased stability, enhanced conductivity, and improved catalytic activity. MOFs can undergo high temperature treatment to produce MOF-derived materials with unique properties. This process is also known as pyrolysis, and it involves heating the MOF at high temperatures in an inert atmosphere or vacuum. An important category of MOF-derived material obtained through pyrolysis is carbonaceous materials. By pyrolyzing MOFs that contain organic ligands, the organic ligands can be carbonized to produce porous carbon materials. These carbonaceous materials can have high surface areas, tunable pore sizes, and excellent electrical conductivity. Another example of MOF-derived material obtained through pyrolysis is metal-containing materials. By pyrolyzing MOFs that contain metal ions, the metal ions can be reduced to form metal nanoparticles or metal carbides. These metal-containing materials can have unique catalytic or magnetic properties (Tables 1.2 and 1.3).

1.7 PROPERTIES OF MOFs

The properties of MOFs are influenced by their chemical composition, structure, and porosity and can be tailored for specific applications. Here are some of the important properties of MOFs:

TABLE 1.3
MOF-derived materials and applications

MOF derivative	MOF precursor	Application	Ref.
N-doped porous carbon	ZIF-8	Battery	[21]
N-doped porous carbon	ZIF-7	Oxygen reduction reaction	[22]
Co-Al LDH@Fe_2O_3	Co-MOF	Adsorption	[23]
Porous carbon	Zn-MOF	Gas storage	[24]
Porous Co_3O_4	Co-MOF	Supercapacitor	[25]
CoNiP	CoNi-MOF	Hydrogen evolution reaction	[26]
NiCo-S	NiCo-MOF-74	Supercapacitor	[27]
Co@NC/MWCNT	ZIF-67	Sensing	[28]
In_2O_3-NiO	In-Ni MOF	Sensing	[29]
FeNi-P(O)	FeNi-MOF	Water splitting	[30]

    a.   Porosity and surface area

       MOFs are a unique class of crystalline porous materials. They possess permanent porosity and high surface area due to the strong bond between metal ions and organic ligands. Porosity can be imparted in MOFs by tuning the inorganic and organic building blocks. Moreover, MOFs possess a high degree of porosity, which allows them to adsorb and store gases, liquids, and small molecules. The porosity of MOFs can be tuned by adjusting the size and shape of the pores, making them ideal materials for gas storage, separation, and catalysis. MOFs showing permanent porosity was first reported in 1998 by Yaghi, i.e. Zn(BDC) MOF. In order to study the permanent porosity of MOF, physisorption isotherms obtained using probe molecules N_2 or Ar at their standard boiling points can be used. Experimental gas sorption measurements in the de-solvated state of MOF using nitrogen adsorption/desorption isotherm at 77 K and low pressure provide information regarding the surface area, pore volume, and pore size distribution. Brunauer-Emmett-Teller (BET) method is commonly used to calculate the surface area, whereas density functional theory (DFT), Monte Carlo simulation, and Horvath–Kawazoe (HK) methods were developed over the years to evaluate the pore size distribution in MOFs. In addition, MOFs have a high surface area due to their porous structure, which provides a large number of active sites for chemical reactions. The surface area of MOFs can be further increased by post-synthetic modification or by adding functional groups to the organic ligands, which can enhance their catalytic activity.

       Along with design and synthesis strategies, activation step also play an important role in achieving higher porosity and stability in MOFs. The as-synthesized MOF often contains guest species such as solvent, metal clusters, or unreacted ligands. In order to remove the guest species from pores and thereby increase the accessibility of pore space and surface area, suitable activation methods should be adopted. The different activation methods used in

MOF synthesis are (i) thermal activation. It is a widely used conventional method suitable for the activation of chemically and thermally stable MOFs. This involves the pre-treatment of synthesized MOF by solvent exchange and the use of vacuum as additional requirement; (ii) solvent exchange which involves the exchange of high boiling point and/or high surface tension solvents used in the synthesis and that occupying the pores with low boiling point and/or low surface tension solvents; and (iii) supercritical CO_2 drying. This consists of three steps: exchanging the solvents in synthesis with organic solvents which are compatible with liquid CO_2, replacing organic solvents with CO_2, and evaporating CO_2 in the supercritical state.

b. Chemical stability

The term "stability" of a material indicates the extent of structural integrity or the resistance of a structure to degradation in harsh environment. MOFs exhibit high chemical stability, making them resistant to harsh chemical environments and high temperatures. This property is due to the strong coordination bonds between the metal ions and the organic ligands. However, some MOFs may be susceptible to hydrolysis or oxidation under certain conditions. Therefore, chemical stability of an MOF includes its stability in acidic, alkaline, and neutral medium. In particular, the structural component in MOF, i.e. metal-linker bond, which is prone to various interactions, decides the stability of the framework. In acidic medium, MOFs undergo hydrolysis to yield protonated linker, and in basic solution, hydroxide ligated nodes are generated.

Basically, functionally stable MOFs are generated from a strong association between the metal node and the ligand. There are different ways of designing stable MOFs which mainly include the combination of metal and ligand units according to the hard and soft acids and bases (HSAB) principle. For example, (i) combination of high valent metal ions (hard acids) with oxygen-containing (carboxylate) ligands (hard bases) yields acid stable MOFs. The archetypal Zr-MOF reported in 2008, termed as UiO-66, containing Zr(IV) nodes and benzene dicarboxylic (BDC) linker is a benchmark MOF which is water stable. Furthermore, Material Institute of Lavoisier (MIL) compounds mostly containing BDC linker and trivalent metal ions like Cr^{3+}, Al^{3+}, and Fe^{3+} exhibit excellent water stability. (ii) Combination of low valent metal ions (soft acids) with nitrogen-containing (azolate) ligands (soft bases) generate alkali-stable MOFs. The strength of a particular metal-nitrogen bond in MOF depends on the pKa value of corresponding azolate ligand deprotonation. Higher the pKa value, higher will be the stability of synthesized MOF. For example, H_3BTP ligand with pKa 19.8 forms hydrolytically stable MOF. (iii) Increase the hydrophobicity of the framework by using organic linkers with hydrophobic substituents such as $-CH_3$ or $-CF_3$ groups. Along with bond strength, shielding effect of substituents is also an important factor in stabilizing MOF. In the case of MOF with more open metal sites, shielding is effective since it can reduce the pore volume to access water molecules. (iv) Substitution of labile building blocks with stable ones through metal metathesis to obtain stable

MOFs. In addition, the dispersion interaction between frameworks, connectivity between building units, rigidity of ligands, geometry of building units, etc. must be taken into account for the design of chemically stable MOFs.

c. Thermal stability

MOFs are generally thermally stable, but some may undergo thermal degradation or decomposition at high temperatures, which can limit their applications in certain fields. Similar to chemical stability, thermal stability of an MOF is related to the strength of metal-linker bond. Furthermore, the number of linkers connected to each metal node also plays an important role in deciding the thermal stability. Typically, thermal decomposition of an MOF generally refers to the breakage of metal-ligand bond associated with combustion of organic ligands. In order to enhance the thermal stability of MOF, the strength of metal-ligand bond has to be increased. This can be achieved by combining metal and ligand units according to hard acid-hard base interaction, changing the composition of ligand pendant groups and framework catenation. Framework catenation is a strategy to increase the stability of MOF through framework-framework interactions arising from the interpenetration of frameworks.

The different analytical techniques used to study the thermal stability of MOF includes thermogravimetric analysis (TGA), differential scanning calorimetry (DSC), in situ powder X-Ray diffraction (PXRD), etc. TGA performed under N_2 atmosphere gives the thermal profile and information regarding the decomposition temperature of the framework. The voids of MOFs are occupied with guest (solvent) molecules used in the synthesis which can be removed in the 100°C–200°C region, followed by a plateau and complete at framework decomposition. TGA is useful to understand the decomposition temperature, whereas DSC can be used to study the phase transitions in MOF. In situ PXRD gives detailed information regarding the structural changes in the framework.

d. Flexibility

The flexibility of MOFs is a unique property that sets them apart from other porous materials and has led to a range of potential applications such as gas storage, separation, and sensing. One of the key advantages of MOFs is their ability to undergo structural transformations in response to changes in their environment. These transformations can be induced by a variety of stimuli, including temperature, pressure, and guest molecules. The flexibility of MOFs allows them to adapt to their environment, which can enhance their properties and functionality. The flexibility of MOFs arises from the dynamic nature of the metal-ligand coordination bonds. These bonds can be broken and re-formed in response to external stimuli, resulting in changes in the overall structure of MOFs. For example, when MOFs are exposed to adsorbates, such as gases or liquids, the metal-ligand coordination bonds can rearrange to accommodate the adsorbates, leading to the expansion of the MOF structure. This expansion can increase the surface area of MOFs and enhance their gas storage and separation properties.

Another example of the flexibility of MOFs is their ability to undergo gate-opening and closing transitions. These transitions involve the opening and closing of pores in response to changes in their environment, such as the presence of guest molecules. This property makes MOFs highly attractive for gas separation and storage applications, as they can selectively adsorb and release gases based on their size and polarity. The structural changes of MOFs in response to external stimuli can also be exploited for sensing applications, where the changes in the MOF structure can be used to detect the presence of specific analytes. Despite the potential benefits of MOF flexibility, there are also challenges associated with this property. For example, the structural changes of MOFs can be irreversible or result in the loss of stability, which can limit their applications. There is also a need for better understanding of the mechanisms of MOF flexibility, as well as the development of strategies for controlling and optimizing this property for specific applications.

1.8 FUNCTIONALIZATION OF MOFs

One important aspect of MOFs is their ability to be functionalized, or modified, with different chemical groups to impart new or improved properties. Functionalization of MOFs can be achieved by either modifying the metal node or ligand, which involves changing the metal ion or cluster coordinated to ligand or modifying the ligand structure. The choice of approach depends on the specific functional group to be introduced and the nature of the MOF and its ligands. This modification can lead to changes in the electronic properties of MOF, as well as changes in its structural and chemical properties. One approach to modify the metal node is to replace the metal ion or cluster with a different one. This can be achieved through a process called "metathesis", in which the metal ion or cluster is exchanged with another one. Metathesis can be carried out by reacting the MOF with a solution containing the desired metal ion or cluster.

Another approach to functionalize MOF is by introducing new functional groups to the ligands coordinated to the metal ion or cluster. This involves post-synthetic modification (PSM), pre-synthetic modification (PrSM), and in-situ modification. PSM consists of addition of functional groups to the ligands after the MOF has been synthesized. PSM can be carried out by a variety of chemical reactions, such as amidation, esterification, and alkylation. The choice of reaction depends on the nature of the ligands and the functional group to be introduced. In short, PSM involves the modification of a pre-existing MOF through chemical reactions. Secondly, PrSM involves introducing functional groups to the ligands before the synthesis of MOF. This can be achieved by synthesizing modified ligands with the desired functional groups and then using them to construct the MOF. In the third case, in-situ modification is achieved by introducing functional groups during MOF growth. This can be achieved by adding a functionalized ligand to the MOF synthesis mixture, which can then coordinate with the metal ion or cluster during MOF growth, i.e., in-situ modification involves the addition of functional groups during MOF growth.

The different types of functional groups that can be added to MOFs include amino group, carboxylic acid group, and phosphonate group. It is very important to select the appropriate functional group based on the desired application and the properties of the MOF. Functionalization has an impact on the properties of MOFs, including their stability, surface area, and selectivity. Furthermore, it is very critical to maintain the structural integrity of the MOF during functionalization and the need for careful characterization to ensure the desired properties are achieved. In conclusion, the functionalization of MOFs is an important aspect of their design and application. Through the addition of different functional groups, MOFs can be tailored to specific applications and their properties can be improved.

1.9 CONCLUSION

MOFs are a relatively young class of materials that have gained increasing attention over the past few decades due to their unique properties and potential applications. The history of MOF research is characterized by a rapid pace of discovery and innovation, driven by its excellent properties. This chapter provides an overview of MOFs, including their synthesis, structural features, and properties. The chemistry of metal-organic frameworks (MOFs) is centred on their unique structure and properties, which result from the combination of metal ions or clusters and organic linkers. The chemistry of MOFs can be divided into several key areas, including synthesis, structure, and reactivity. The choice of metal and linker can influence the properties of the resulting MOF, including its pore size, surface area, and reactivity. Furthermore, the structure of MOFs is determined by the coordination geometry of the metal ions or clusters and the topology of the organic linkers. Another crucial parameter is the reactivity, which is governed by the properties of the metal ions or clusters and the organic linkers. MOFs can exhibit a range of chemical and physical properties, including catalytic activity, gas adsorption and separation, and drug delivery. The reactivity of MOFs can be modified through the introduction of functional groups or the use of different metal ions or linkers.

The synthesis and characterization of MOFs have been the subject of extensive research in recent years, with a range of methods available for the production of MOFs with specific properties. The unique properties of MOFs make them attractive candidates for a wide range of applications in fields such as energy, catalysis, and medicine. Further research is needed to fully understand and optimize the properties of MOFs, but their potential for use in a range of applications makes them an exciting area of research for materials scientists and engineers.

REFERENCES

1.   Yaghi, O.; Li, H. Hydrothermal Synthesis of a Metal-Organic Framework Containing Large Rectangular Channels. *J Am Chem Soc*, 1995, *117* (41), 10401–10402.
2.   Li, H.; Eddaoudi, M.; O'Keeffe, M.; Yaghi, O. M. Design and Synthesis of an Exceptionally Stable and Highly Porous Metal-Organic Framework. *Nature*, 1999, *402* (6759), 276–279.

3. Cai, M.; Ni, B.; Hu, X.; Wang, K.; Yin, D.; Chen, G.; Fu, T.; Zhu, R.; Dong, X.; Qu, C. An Investigation of IRMOF-16 as a pH-Responsive Drug Delivery Carrier of Curcumin. *J Sci: Adv Mater Devices*, 2022, *7* (4), 100507.

4. Chui, S. S.-Y.; Lo, S. M.-F.; Charmant, J. P.; Orpen, A. G.; Williams, I. D. A Chemically Functionalizable Nanoporous Material [Cu3(TMA)2 (H2O)3]n. *Science*, 1999, *283* (5405), 1148–1150.

5. Cavka, J. H.; Jakobsen, S.; Olsbye, U.; Guillou, N.; Lamberti, C.; Bordiga, S.; Lillerud, K. P. A New Zirconium Inorganic Building Brick Forming Metal Organic Frameworks with Exceptional Stability. *J Am Chem Soc*, 2008, *130* (42), 13850–13851.

6. Ye, X.; Liu, D. Metal–Organic Framework UiO-68 and Its Derivatives with Sufficiently Good Properties and Performance Show Promising Prospects in Potential Industrial Applications. *Cryst Growth Des*, 2021, *21* (8), 4780–4804.

7. Park, K. S.; Ni, Z.; Côté, A. P.; Choi, J. Y.; Huang, R.; Uribe-Romo, F. J.; Chae, H. K.; O'Keeffe, M.; Yaghi, O. M. Exceptional Chemical and Thermal Stability of Zeolitic Imidazolate Frameworks. *Proc Natl Acad Sci*, 2006, *103* (27), 10186–10191.

8. Morris, W.; Doonan, C. J.; Furukawa, H.; Banerjee, R.; Yaghi, O. M. Crystals as Molecules: Postsynthesis Covalent Functionalization of Zeolitic Imidazolate Frameworks. *J Am Chem Soc*, 2008, *130* (38), 12626–12627.

9. Millange, F.; Serre, C.; Férey, G. Synthesis, Structure Determination and Properties of MIL-53as and MIL-53ht: The First Cr iii Hybrid Inorganic–Organic Microporous Solids: Cr iii (OH)·{O2C–C6H4–CO2}·{HO2C–C6H4–CO2H} x. *Chem Commun*, 2002, (8), 822–823.

10. Patil, D. V.; Rallapalli, P. B. S.; Dangi, G. P.; Tayade, R. J.; Somani, R. S.; Bajaj, H. C. MIL-53 (Al): An Efficient Adsorbent for the Removal of Nitrobenzene from Aqueous Solutions. *Ind Eng Chem Res*, 2011, *50* (18), 10516–10524.

11. Li, J.; Xia, J.; Zhang, F.; Wang, Z.; Liu, Q. An Electrochemical Sensor Based on Copper-Based Metal-Organic Frameworks-Graphene Composites for Determination of Dihydroxybenzene Isomers in Water. *Talanta*, 2018, *181*, 80–86.

12. Koo, W.-T.; Jang, J.-S.; Choi, S.-J.; Cho, H.-J.; Kim, I.-D. Metal–Organic Framework Templated Catalysts: Dual Sensitization of PdO–ZnO Composite on Hollow SnO2 Nanotubes for Selective Acetone Sensors. *ACS Appl Mater Interfaces*, 2017, *9* (21), 18069–18077.

13. Yuan, Y.; Jiang, S.; Miao, Q.; Zhang, J.; Wang, M.; An, L.; Cao, Q.; Guan, Y.; Zhang, Q.; Liang, G. Fluorescent Switch for Fast and Selective Detection of Mercury (II) Ions In Vitro and in Living Cells and a Simple Device for Its Removal. *Talanta*, 2014, *125*, 204–209.

14. Liu, X.; Yan, Z.; Zhang, Y.; Liu, Z.; Sun, Y.; Ren, J.; Qu, X. Two-Dimensional Metal–Organic Framework/Enzyme Hybrid Nanocatalyst as a Benign and Self-Activated Cascade Reagent for In Vivo Wound Healing. *Acs Nano*, 2019, *13* (5), 5222–5230.

15. Cho, K.; Han, S. H.; Suh, M. P. Copper–Organic Framework Fabricated with CuS Nanoparticles: Synthesis, Electrical Conductivity, and Electrocatalytic Activities for Oxygen Reduction Reaction. *Angew Chem*, 2016, *128* (49), 15527–15531.

16. Chen, B.; Li, Y.; Li, M.; Cui, M.; Xu, W.; Li, L.; Sun, Y.; Wang, M.; Zhang, Y.; Chen, K. Rapid Adsorption of Tetracycline in Aqueous Solution by Using MOF-525/Graphene Oxide Composite. *Microporous Mesoporous Mater*, 2021, *328*, 111457.

17. Jayaramulu, K.; Dubal, D. P.; Schneemann, A.; Ranc, V.; Perez-Reyes, C.; Stráská, J.; Kment, Š.; Otyepka, M.; Fischer, R. A.; Zbořil, R. Shape-Assisted 2D MOF/Graphene Derived Hybrids as Exceptional Lithium-Ion Battery Electrodes. *Adv Funct Mater*, 2019, *29* (38), 1902539.

18. Ge, L.; Zhou, W.; Rudolph, V.; Zhu, Z. Mixed Matrix Membranes Incorporated with Size-Reduced Cu-BTC for Improved Gas Separation. *J Mater Chem A*, 2013, *1* (21), 6350–6358.

19. Brahmi, C.; Benltifa, M.; Vaulot, C.; Michelin, L.; Dumur, F.; Millange, F.; Frigoli, M.; Airoudj, A.; Morlet-Savary, F.; Bousselmi, L. New Hybrid MOF/Polymer Composites for the Photodegradation of Organic Dyes. *Eur Polym J*, 2021, *154*, 110560.

20. Ke, F.; Yuan, Y.-P.; Qiu, L.-G.; Shen, Y.-H.; Xie, A.-J.; Zhu, J.-F.; Tian, X.-Y.; Zhang, L.-D. Facile Fabrication of Magnetic Metal–Organic Framework Nanocomposites for Potential Targeted Drug Delivery. *J Mater Chem*, 2011, *21* (11), 3843–3848.

21. Zheng, F.; Yang, Y.; Chen, Q. High Lithium Anodic Performance of Highly Nitrogen-Doped Porous Carbon Prepared from a Metal-Organic Framework. *Nat Commun*, 2014, *5* (1), 5261.

22. Zhang, P.; Sun, F.; Xiang, Z.; Shen, Z.; Yun, J.; Cao, D. ZIF-Derived In Situ Nitrogen-Doped Porous Carbons as Efficient Metal-Free Electrocatalysts for Oxygen Reduction Reaction. *Energy Environ Sci*, 2014, *7* (1), 442–450.

23. Poudel, M. B.; Awasthi, G. P.; Kim, H. J. Novel Insight into the Adsorption of Cr (VI) and Pb (II) Ions by MOF Derived Co-Al Layered Double Hydroxide@ Hematite Nanorods on 3D Porous Carbon Nanofiber Network. *Chem Eng J*, 2021, *417*, 129312.

24. Chen, S.; Li, Y.; Mi, L. Porous Carbon Derived from Metal Organic Framework for Gas Storage and Separation: The Size Effect. *Inorg Chem Commun*, 2020, *118*, 107999.

25. Li, G.-C.; Hua, X.-N.; Liu, P.-F.; Xie, Y.-X.; Han, L. Porous Co3O4 Microflowers Prepared by Thermolysis of Metal-Organic Framework for Supercapacitor. *Mater Chem Phys*, 2015, *168*, 127–131.

26. Lu, Y.; Deng, Y.; Lu, S.; Liu, Y.; Lang, J.; Cao, X.; Gu, H. MOF-Derived Cobalt–Nickel Phosphide Nanoboxes as Electrocatalysts for the Hydrogen Evolution Reaction. *Nanoscale*, 2019, *11* (44), 21259–21265.

27. Yu, J.; Gao, X.; Cui, Z.; Jiao, Y.; Zhang, Q.; Dong, H.; Yu, L.; Dong, L. Facile Synthesis of Binary Transition Metal Sulfide Tubes Derived from NiCo-MOF-74 for High-Performance Supercapacitors. *Energy Technol*, 2019, *7* (6), 1900018.

28. Zhao, J.; Long, Y.; He, C.; Yang, H.; Zhao, S.; Luo, X.; Huo, D.; Hou, C. Simultaneous Electrochemical Detection of Cd2+ and Pb2+ Based on an MOF-Derived Carbon Composite Linked with Multiwalled Carbon Nanotubes. *ACS Sustainable Chem Eng*, 2023, *11*, 2160–2171.

29. Yu, Q.; Jin, R.; Zhao, L.; Wang, T.; Liu, F.; Yan, X.; Wang, C.; Sun, P.; Lu, G. MOF-Derived Mesoporous and Hierarchical Hollow-Structured In2O3-NiO Composites for Enhanced Triethylamine Sensing. *ACS Sens*, 2021, *6* (9), 3451–3461.

30. Lin, C.; Wang, D.; Jin, H.; Wang, P.; Chen, D.; Liu, B.; Mu, S. Construction of an Iron and Oxygen Co-Doped Nickel Phosphide Based on MOF Derivatives for Highly Efficient and Long-Enduring Water Splitting. *J Mater Chem A*, 2020, *8* (8), 4570–4578.

2 Metal–Organic Framework Derived Materials
Classification and Synthesis

2.1 INTRODUCTION

Even though MOFs possess a highly ordered crystalline structure with excellent properties like large surface area, tunable porosity, and presence of coordinatively unsaturated sites/open metal sites, the inferior electrical conductivity and poor chemical stability in aqueous medium hinder its application in research areas. In the broad spectrum of MOF-based materials, the class of MOF-derived materials, or commonly known as MOF derivatives, holds a significant place since it can address the innate drawbacks of pristine MOF and its composites. This aspect widens the scope of application of MOF-based materials mainly in the field of electrochemistry. The class of MOF-derived materials include MOF-derived carbon materials, metallic species, and their composites which are obtained by thermal, chemical, or electrochemical decomposition of MOFs under controlled conditions. The resulting materials inherit some of the structural features of the parent MOF but with modified properties and enhanced functionality. Furthermore, the morphology, size, and other features of MOF can be well preserved in the derived material by incorporation of rational design and synthesis methods. Generally, they possess higher chemical stability, electrical conductivity, and electrochemical activity compared to pristine MOF and MOF-based composites which is favourable for its application in catalysis, adsorption, sensing, energy storage, and conversion. The versatility of MOF-derived materials arises from the wide range of metal ions and organic ligands used to construct the parent MOF. By carefully selecting the components, researchers can tailor the properties of the MOF and subsequently control the characteristics of the derived materials (Figure 2.1).

2.2 TIMELINE OF MOF-DERIVED MATERIALS

The history and timeline of MOF-derived materials trace back to the discovery and development of MOFs as a distinct class of porous materials. Here is a chronological overview of the key milestones and advancements in the field. Among the class of MOF materials, the section of MOF derivatives arrived later into the picture.

DOI: 10.1201/9781003432357-3

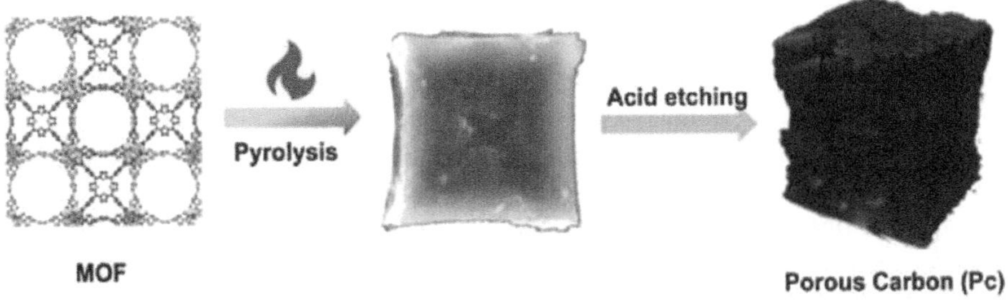

FIGURE 2.1 Schematic of the use of MOF as sacrificial template to prepare a porous carbon. Reproduced with permission from Ref. [1].

2004: The first example of using MOFs as precursors for the synthesis of porous carbon materials was reported by Yaghi and colleagues. The MOF-177 precursor was carbonized to obtain a high-surface-area carbon material, demonstrating the potential for MOF-derived materials.

2005: MOF-derived materials started gaining attention as potential catalysts. Researchers demonstrated the conversion of MOF-5 into catalytically active metal nanoparticles, such as palladium and platinum, for various reactions.

2008: A breakthrough in MOF-derived materials came with the work of Yaghi and colleagues, who developed a general strategy for the synthesis of MOF-derived carbon materials. By carbonizing different MOFs, they obtained porous carbon materials with tailored porosity and surface chemistry.

2010s: The field witnessed a surge in research on MOF-derived materials. Researchers explored different methods, such as pyrolysis, chemical etching, and template removal, to obtain carbon-based and metal-based materials from MOF precursors.

2012: MOF-derived materials found applications in energy storage. Researchers reported the synthesis of porous carbon materials with hierarchical structure from MOFs for supercapacitors and lithium-ion batteries, demonstrating their potential in energy storage devices.

2014: The concept of MOF-derived composites gained attention. Researchers combined MOF-derived carbon materials with other components, such as polymers or nanoparticles, to create hybrid materials with enhanced properties for applications in catalysis, sensing, and biomedicine.

2017: The field expanded to include MOF-derived metal oxides. Researchers explored the conversion of MOFs into metal oxide materials through controlled thermal or chemical processes, leading to applications in catalysis, energy storage, and gas sensing.

2019: Advances in MOF-derived materials included the synthesis of MOF-derived thin films. Researchers developed methods to deposit MOF precursors onto substrates and subsequently convert them into thin films, enabling applications in sensors, membranes, and electronics.

2020s: The research focus on MOF-derived materials broadened to include functionalization and post-synthesis modifications. Researchers explored the incorporation of certain functional groups or molecules into MOF-derived materials to enhance their properties and enable targeted applications.

The history and timeline of MOF-derived materials highlight their evolution from the discovery of MOFs to the development of various synthesis strategies and applications. The field continues to advance, driven by the need for advanced materials with tailored properties and functionalities for diverse scientific and technological applications.

2.3 CLASSIFICATION

Classification of MOF-derived materials involves a comprehensive understanding of the various synthesis strategies, structural modifications, and applications of these materials. The classification can be based on multiple factors, including precursor selection, activation methods, structural characteristics, and functional properties (Figure 2.2).

   a. Porous carbons: MOF-derived porous carbons are obtained by the carbonization of precursor MOFs. The resulting materials possess high surface area, tunable porosity, and excellent adsorption properties. They find applications in gas storage, energy storage, and catalysis.
   b. Carbon nanomaterials: MOF-derived carbon nanomaterials include carbon nanotubes [2], graphene, and graphene-like structures. These materials exhibit exceptional mechanical, electrical, and thermal properties, making them suitable for applications in electronics, energy storage, and composites.

Metal and metal oxide MOF-derived materials:

   a. Metal nanoparticles: MOF-derived materials can be transformed into metal nanoparticles through controlled thermal or chemical reduction. These metal nanoparticles can exhibit unique catalytic, optical, or magnetic properties, leading to applications in catalysis, sensors, and nanomedicine.

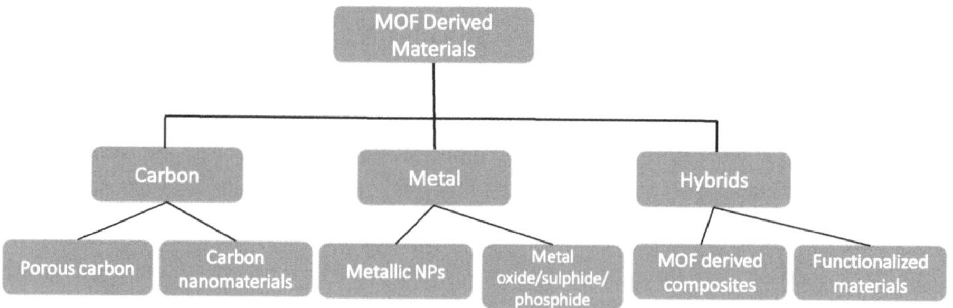

FIGURE 2.2 Classification of MOF-derived materials.

b. Metal oxides: MOF-derived metal oxides are obtained by the controlled oxidation or thermal decomposition of MOFs. These materials possess tailored compositions and structures, making them suitable for applications in energy storage, photocatalysis, and gas sensing.

Hybrid MOF-derived materials:

a. MOF-derived composites: MOF-derived materials can be combined with other components, such as polymers, nanoparticles, or graphene, to form composite materials. These composites exhibit synergistic properties, combining the advantages of MOFs and the added components. They find applications in energy storage, drug delivery, and environmental remediation.
b. Functionalized MOF-derived materials: MOF-derived materials can be further functionalized by incorporating specific functional groups or molecules during the synthesis or post-synthesis modification. This allows for the introduction of specific properties, such as enhanced catalytic activity, selectivity, or biocompatibility.

Morphological classification:
MOF-derived materials can also be classified based on their morphology or structure:

a. Nanoparticles: MOF-derived nanoparticles exhibit size-dependent properties and find applications in catalysis, imaging, and drug delivery.
b. Thin films: MOF-derived thin films can be obtained through various deposition techniques, offering controlled surface properties and applications in sensing, membranes, and electronics.
c. Hierarchical structures: MOF-derived materials with hierarchical structures possess multiple levels of porosity, enhancing mass transport and accessibility to active sites which can be utilized in adsorption, catalysis, and energy storage.

It is important to note that the classification of MOF-derived materials is not mutually exclusive, and many materials can fall into multiple categories depending on their synthesis approach, modifications, or applications. The field of MOF-derived materials is constantly evolving, and further research and exploration are expected to expand the classification and uncover new categories and applications for these versatile materials.

2.4 STRUCTURE

The structure of MOF-derived materials is influenced by the nature of precursor MOF, the activation process, and any subsequent modifications or functionalization. MOF-derived materials can exhibit diverse structures, ranging from amorphous carbon to crystalline carbon frameworks, metal nanoparticles, or metal oxide structures. Here are some common structural characteristics of MOF-derived materials:

Amorphous carbon: MOF-derived materials can have an amorphous carbon structure, lacking long-range order or crystallinity. Amorphous carbon materials often possess a disordered arrangement of carbon atoms, resulting in a high surface area and a range of pore sizes. The specific carbon structure can vary depending on the precursor MOF and the carbonization conditions.

Crystalline carbon frameworks: Some MOF-derived materials can exhibit crystalline carbon frameworks, resembling the parent MOF structure. These materials retain the porosity and network connectivity of the precursor MOF, while the organic ligands are carbonized or removed during the activation process. Crystalline carbon frameworks can have well-defined pore sizes and surface chemistry, allowing for efficient adsorption and catalytic reactions.

Metal nanoparticles: MOF-derived materials can involve the formation of metal nanoparticles dispersed within a carbon or oxide matrix. The metal nanoparticles can range in size from a few nanometres to tens of nanometres and can exhibit different shapes and distributions. The metal nanoparticles can be derived from the metal nodes present in the original MOF structure or can be introduced during the carbonization process.

Metal oxide structures: In some cases, MOF-derived materials can undergo controlled thermal or chemical treatments to yield metal oxide structures. The metal oxide materials inherit the morphology and pore structure of the precursor MOF but with the metal centres converted into metal oxides. These metal oxide structures can exhibit specific surface chemistry and functionalities suitable for energy storage, catalysis, and sensing applications.

Composite structures: MOF-derived materials can also be designed as composites, combining the properties of MOFs with other components. For example, the combination of MOF-derived carbon materials with polymers, nanoparticles, or graphene creates composites with tailored properties. These composites can have unique structures, for example, core-shell architectures or hierarchical arrangements, resulting in enhanced functionalities and performance.

It is important to note that the structure of MOF-derived materials can be influenced by the specific synthesis and activation methods employed. Different carbonization temperatures, activation atmospheres, and post-synthesis treatments can lead to variations in the final structure and physico-chemical properties of the derived materials. Researchers continue to explore novel synthesis strategies and modifications to tailor the structure of MOF-derived materials for specific applications.

2.5　MOF-DERIVED CARBON MATERIALS

The distinct advantage of MOFs lies in their ability to modulate composition and structure by judicious selection of organic ligands and metal components. This imparts MOF materials with a diverse composition and structure, an assortment of pore architectures, and exceptional porosity, collectively rendering them as excellent templates and precursors for the synthesis of porous carbon materials. Carbon materials derived from MOFs adeptly preserve the pronounced pore architecture and extensive specific surface area inherent to their precursor MOFs, while also demonstrating notable electrical conductivity and robust stability [3]. Through deliberate selection of precursor MOF, the resulting carbon materials can be modified via in situ heteroatom doping (e.g., N, P, S, B) characterized by uniform and well-ordered atomic dispersion profiles. Furthermore, the specific surface area, morphology, and particle dimensions of MOF-derived carbon materials can be methodically tailored via meticulous synthetic control, underscoring their position as a competitive and versatile category of carbonaceous materials.

Porous carbons are commonly acquired by carbonizing precursors sourced from nature or synthesized materials, succeeded by activation. Diverse approaches have been investigated to craft porous carbons with regulated pore architectures, spanning both micropores and mesopores. Specifically, in terms of micropores, molecular sieving carbons with consistent micropore sizes can be synthesized through the carbonization of well-suited carbon precursors. In addition, carbon materials possessing mesopores, formed through catalytic activation and subsequent carbonization, often tend to exhibit irregular arrangements and broad pore size distributions. This outcome stems from the etching processes responsible for creating pores in these disordered porous carbons, which are challenging to regulate, or from gases escaping during the carbonization of carbon precursors.

To enhance structural coherence, the application of a template can direct pore formation during the carbonization procedure. By adopting template carbonization methods, it becomes feasible to fabricate carbon materials with controlled structures and relatively narrower pore size distributions.

2.5.1 Porous Carbon Materials – Overview

Nanoporous carbons stand out as a significant class of porous materials, valued for their robust thermal and chemical stability. These materials have conventionally been synthesized through pyrolysis, followed by either chemical or physical activation of organic component. Despite resulting in activated porous carbons with extensive surface areas, these structures exhibit disorderliness, featuring wide-ranging pore size distributions. Consequently, their practical applications in molecular discrimination are limited. To address this limitation, efforts have been directed towards creating nanoporous carbons with ordered pore structures and narrower size distributions, employing templating techniques.

One approach involves the use of ordered hard-templates, such as nanocasts or hard-templated carbon. This method encompasses carbonization or chemical vapour deposition of carbon sources within these templates. Nanocasting technique is an efficient method to prepare micro- and mesoporous carbons, where zeolite, layered double hydroxide (LDH), SBA-15, etc., can be used as porous templates and furfuryl alcohol (FA) [4], urea, sucrose [5], etc., are used as carbon precursors. Alternatively, ordered polymeric carbon gels, generated from soft-templating using block copolymers and resols, can be carbonized to produce what is known as soft-templated carbon. This latter technique yields nanoporous carbons exhibiting diverse structures and morphologies. However, its success relies on the combination of thermally stable carbon gels with organic templates that decompose before the carbonization process completes.

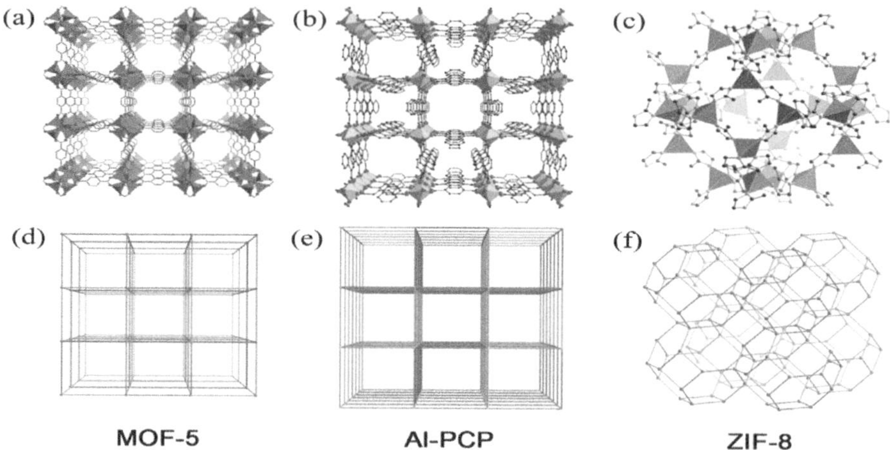

FIGURE 2.3   Some representative MOFs used to create porous carbon by sacrificial template method. Reproduced with permission from Ref. [6].

Furthermore, the nanocasting method offers a range of ordered nanoporous carbons, with varying pore sizes (micro-, meso-, and macropores), dependent on the initial hard-templates employed, such as zeolites, mesoporous silicas, and colloidal crystals. It is worth noting that this nanocasting method, while effective in achieving ordered nanoporous structures, is slightly intricate and not the most suitable choice for large-scale production (Figure 2.3).

The synthesis of MOF-derived nanoporous carbons involves multiple steps to transform the original MOF structure into a carbonaceous material with desirable properties. The following methods have been employed in this process.

2.5.2   Template Removal and Carbonization

MOFs have become potential candidates for the use of template and precursors for porous carbon owing to their large surface area, tunable structure, adjustable porosity, and high organic content. Porous MOFs possess inherent nanoscale cavities and accessible channels that create favourable environments for the ingress of small molecules. This characteristic makes them promising candidates for serving as templates in the development of nanoporous carbon materials [7]. In the early stages of research, utilizing MOFs as the template and FA as the carbon precursor allows for the creation of porous carbons [8]. Furthermore, the use of MOF as both template and carbon precursor gained more interest. Here, along with MOF, some additional carbon sources were also added such as carbon tetrachloride, phenolic resin, and ethylenediamine. In this method, the initial step involves the removal of the templating species within the MOF framework, often organic ligands or solvent molecules. This is followed by controlled pyrolysis or carbonization under an inert atmosphere, which thermally decomposes the organic part, leaving behind a carbon-based structure with preserved porosity. The different carbon sources can result in diverse carbon structures (Figures 2.4 and 2.5).

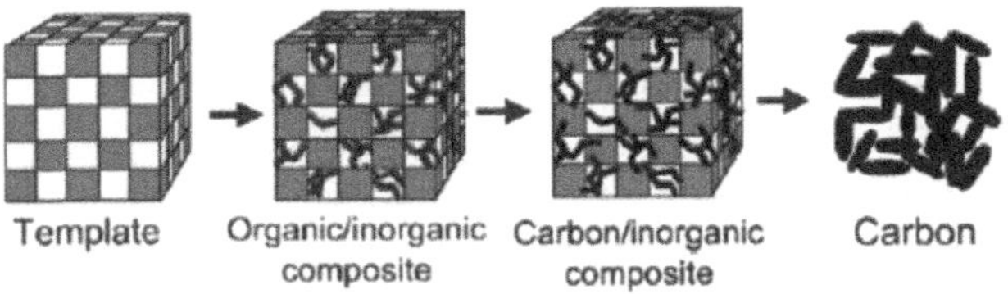

FIGURE 2.4 Schematic of the synthesis of porous carbon by template removal and carbonization. Reproduced with permission from Ref. [9].

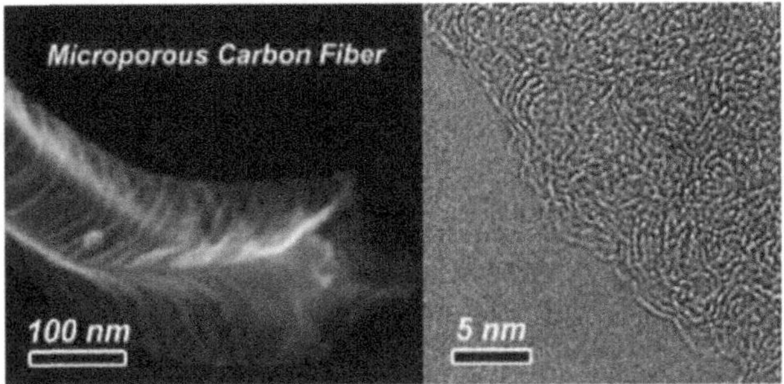

FIGURE 2.5 FESEM and TEM images of microporous carbon fibre. Reproduced with permission from Ref. [4].

2.5.3 DIRECT CARBONIZATION OF MOF PRECURSOR

The most widely used synthesis strategy for porous carbon is the direct conversion of MOF which is found to be the simple and facile route that offers a wide range of MOF precursors for carbonization. The carbonization parameters such as temperature, heating rate, time, and nature of gas significantly influence the morphology, composition, and structure of porous carbon. By the direct carbonization method, MOFs can be easily converted into carbon-based materials by a process called pyrolysis in an inert atmosphere under optimum conditions. This strategy imparts suitable microstructure and surface properties to the derived porous carbon by the careful modulation of temperature and post-carbonization treatment. This also allows for the transfer of characteristics of MOF pore structure into the porous carbon. Meanwhile, the composition of MOF-derived carbon can be controlled by the doping of heteroatoms into the carbon framework and substitution of metal ions in the secondary building unit. The techniques for the control of composition and morphology of derived carbon are discussed in this section.

MOFs directly converted into porous carbon at high temperature is the most common and widely used synthesis strategy for the morphology control of derived carbon, where the carbon inherits the framework structure of precursor MOF. Liu et al. reported the first example of using an MOF as template for the preparation of nanoporous carbon, i.e., MOF-5 as template and FA as carbon precursor [7]. The carbonization of PFA/MOF-5 composite was carried out at 1000°C for 8 h under Ar atmosphere. Here, the polymerization of FA followed by pyrolysis in the pores of

MOF leads to incomplete filling of pores and results in free space in the framework where carbonization occurs. Eventually, MOF-5 framework decomposes during carbonization and acts as a self-sacrificial template for the formation of nanoporous carbon with porosity ranging from micro to macro having sheet-like structure and high specific surface area. In another example, Guo et al. prepared N and Co-doped porous carbon by the direct carbonization of ZIF-67 [Co(2-methylimidazole)$_2$] at different temperatures (600°C–900°C for 1 h) using ZIF-67 as self-sacrificial template [10]. The near spherical shape of ZIF-67 precursor is retained in the derived carbon obtained at 600°C and 700°C, whereas the metal particles get agglomerated and structure collapsed at 900°C. This clearly indicates that the increase in carbonization temperature significantly changes the morphology of derived carbon materials. Within the realm of MOFs, the zeolitic imidazolate frameworks (ZIF) series stands out as a popular choice for the creation of carbon materials. This preference arises from its facile synthesis and the ability to fine-tune its morphology. In particular, carbon materials derived from ZIF-8 have garnered substantial interest for their high nitrogen content and surface area. Notably, the nitrogen-containing ligand, methylimidazole, serves as a valuable precursor for nitrogen-doped porous carbon synthesis (Figure 2.6 and Table 2.1).

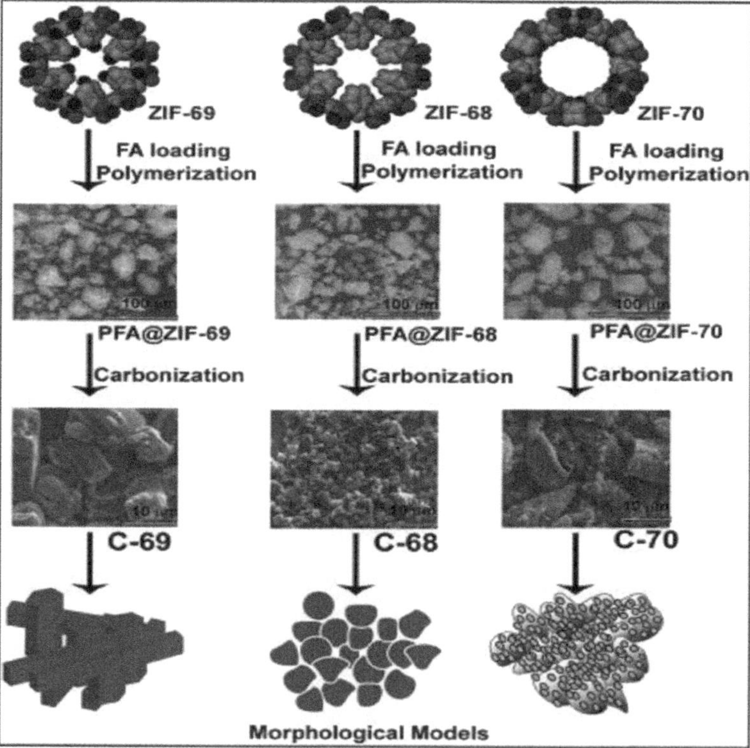

FIGURE 2.6 Synthesis of a series of porous carbons using ZIFs as template at 1000°C. Reproduced with permission from Ref. [12].

TABLE 2.1
Representative examples of porous carbon derived from using MOF as template and direct carbonization

MOF precursor	Role of MOF	Carbon source	Carbonization	Carbon type	Ref.
MOF-5	Template	FA	1000°C, 8 h, Ar	Nanoporous carbon (NPC)	[7]
MOF-5	Template and carbon source	Phenolic resin	200°C (2 h), 600 (4 h) and 900 (2 h)	Porous carbon (MPC)	[11]
Al-PCP	Template	FA	1000°C, 5 h	Microporous carbon	[4]
IRZIF	Template	FA	1000°C	Microporous carbon	[12]
ZIF-8	Direct carbonization	–	600°C–1000°C, Ar	Porous carbon	[13]
MOF-5	Direct carbonization	–	1000°C, 2 h, Ar	Porous carbon	[14]
IRMOF-3	Template	–	200°C (2 h), 600°C–950°C (6 h), Ar	Nitrogen-doped porous carbon	[15]
ZIF-8	Template and carbon source	Melamine, urea, xylitol and sucrose	950°C	Nitrogen-doped porous carbon	[5]
Co@MOF-5	Direct carbonization	–	500°C, 6 h, N_2	Porous carbon	[16]
Ni-MOF-74	Direct carbonization	FA	450°C, 800°C, 1000°C, N_2	Porous carbon	[17]

2.5.4 MOF-Derived Carbon from Matrix-Assisted In Situ Grown MOF

In order to mitigate the issue of particle agglomeration, enhance the specific surface area, and increase electrical conductivity, researchers proposed a method involving the in-situ growth of MOF particles on diverse matrix materials. This technique leads to the formation of composite carbon materials. Nonetheless, there are constraints associated with the size of MOF crystals, and notably, the challenge of preventing MOF crystal breakdown during carbonization. This breakdown poses a hindrance to achieving the continuous electronic conductivity necessary for electrochemical applications [18]. In some cases, carbon materials directly obtained through high-temperature pyrolysis of ZIF nanocrystals inherently suffer from significant structural and morphological disruptions. This includes the formation of discrete carbon particles and a reduced degree of graphitization, ultimately leading to low electrical conductivity and a substantial decline in the number of catalytic active sites. As a potential solution, the approach of anchoring MOF onto matrices such as graphene, graphene oxide (GO), and reduced graphene oxide (rGO) followed by carbonization into matrix-incorporated porous carbon composites presents to be a promising alternative.

TABLE 2.2
Examples of MOF-derived porous carbon from matrix-assisted in-situ-grown MOF

Material	MOF	Matrix	In situ growth method	Pyrolysis conditions	Ref.
Cu@PC	Cu-BTC	Cu foam	Anodic electroprecipitation	850°C, 8 h, Ar	[20]
NiO/C/rGO	Ni-TPA	Graphene oxide	Solvothermal	300°C, 2 h, Air	[21]
LDH@ ZIF-67-800	ZIF-67	Co-Al-LDH	In-situ precipitation	800°C, 2 h, N_2	[22]
RuCo@NC	Ru-Co$_3$[Co (CN)$_6$]$_2$	N-doped graphene	In-situ precipitation	600°C, N_2	[23]
ZIF'-FA-CNT-p	FeZn-MOF	MWCNT	In-situ precipitation	950°C, 2 h, Ar	[24]

For example, Liu et al. prepared sandwich-type structured ZIF-8@GO composite by in situ growth of ZIF-8 nanocrystals on GO surfaces, which was converted into ZIF-8@graphene [19]. Furthermore, N-doped porous carbon@G was obtained by high temperature pyrolysis (900°C, N_2) with well-preserved sheet morphology which possess more catalytic active sites, hierarchical porous structure, and increased graphitization degree suitable for electrocatalytic applications. In another study, Han et al. synthesized a 3D core-shell, ZIF-8 derived carbon@LDH composite, which is characterized by the co-existence of micro-meso-macro pores. Here, ZIF-derived carbon possesses improved conductivity due to the increased accessibility of the active site and accelerated ion migration suitable for supercapacitor application (Table 2.2).

2.5.5 CHEMICAL VAPOUR DEPOSITION (CVD)

A novel structure, comprising a porous carbon framework interwoven with graphitic nanostructures, was crafted through a unique approach by Gadipelli et al. [25]. This method involves utilizing a bimetallic zeolitic imidazolate framework-8 (ZIF-8) as a solid precursor, serving the dual purpose of templating porous carbon and facilitating the growth of graphitic nanocarbon. The decomposition of a ligand (2-methylimidazolate, 2MIM), at temperatures exceeding 600°C, produces active carbon/hydrocarbon radicals and vapours. The central concept underlying the utilization of ZIF-8 revolves around its ability to fulfil multiple roles: (1) serving as a sacrificial template, facilitating the generation of porous carbon through framework decomposition, rearrangement, and the vaporization of zinc metal centres; (2) yielding nitrogen-doped carbon derived from its inherent framework ligand (2MIM), which contains nitrogen; and (3) functioning as a solid precursor to produce carbon radicals during ligand decomposition. These carbon radicals subsequently foster the growth of highly conductive graphitic nanostructures. This growth is further facilitated by the in-situ catalytic properties of transition metal centres within the ZIF-8 framework, all within the framework of a simplified chemical vapour deposition (CVD) process (Figure 2.7).

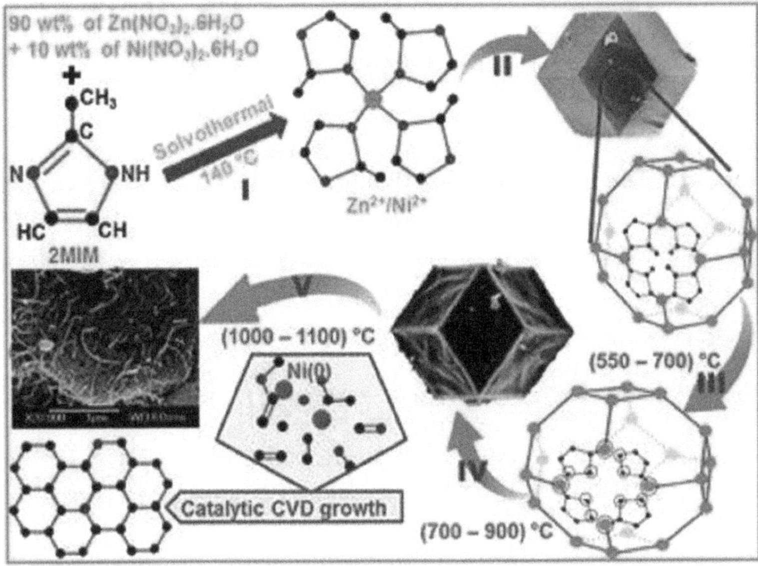

FIGURE 2.7 Schematic representative synthesis of graphitic porous carbon nanostructures. Reproduced with permission from Ref. [25].

2.5.6 DOPING OF HETEROATOM IN THE MOF-DERIVED CARBON

Beyond morphology and graphitization levels, the structural composition of a material plays a pivotal role in influencing its performance. Consequently, regulating the material's composition stands as an effective strategy to optimize its overall performance. The introduction of heteroatoms like nitrogen (N), sulphur (S), boron (B), or phosphorus (P) into the carbon structure represents an effective method for fine-tuning its inherent properties. This includes adjustments to electronic characteristics, surface and local chemical attributes, and mechanical properties. Among these options, nitrogen (N) doping stands out as a particularly appealing choice. This is primarily due to its atomic size closely resembling that of carbon and its possession of five valence electrons. These features facilitate the formation of robust valence bonds with carbon atoms, rendering it highly versatile for a wide array of applications. The process of obtaining N-doped nanoporous carbons (NPCs) involves the direct carbonization of MOF precursors that incorporate nitrogen-containing ligands. Furthermore, in order to enhance the nitrogen content within MOFs, supplementary carbon sources containing nitrogen can be introduced as precursors (Table 2.3).

The porous carbon materials derived from MOFs exhibit a remarkable diversity in their dimensional configurations. These encompass zero-dimensional (0D) structures represented by quantum dots and clusters, one-dimensional (1D) formations embodied by carbon nanotubes, two-dimensional (2D) architectures exemplified by nanosheets with graphene-like characteristics, and three-dimensional (3D) frameworks characterized by extensive porous carbon networks. The crystalline nature and extent of graphitization within these distinct carbon structures can be

TABLE 2.3

Examples of heteroatom and metal-doped porous carbon obtained from carbonization of MOF precursors

Porous carbon	Dopants	Precursor MOF	Pyrolysis conditions	Acid washing	Ref.
Co-doped porous carbon	Co	Co-imidazolate	500°C–900°C, 1 h, Ar	0.5 M H_2SO_4, 2 h	[26]
N- and O-doped porous carbon	N, O	ZIF-8	800°C, 3 h, Ar	1 M HCl	[27]
Co-HPC@CF	Co	ZIF-67@PAN/PVP-1	800°C, 2 h, N_2	–	[28]
Zn-HPC@CF	Zn	ZIF-8@PAN/PVP-1	800°C, 2 h, N_2	–	[28]
Zn/Co-HPC@CF	Zn, Co	BMZIFs@PAN/PVP-1	800°C, 2 h, N_2	–	[28]
3D-BNPCs	B, N, O	ZIF-8	800°C, 8 h, N_2	Con. HCl (36.5%)	[29]
BNPC	B, N	Zn-BPDC	1000°C, 4 h, N_2	2 M HCl	[30]
P-C-600	P	Co-MOF	600°C, N_2	3 M HCl	[31]
BrHT@CoNC	Br, Co, N	ZIF-67	750°C, 2 h, Ar	2 M HCl	[32]

meticulously adjusted. This fine-tuning is achieved by modulating the composition and manipulating processing parameters throughout the pyrolysis procedure.

In 2016, Xia and colleagues introduced an innovative approach for producing nitrogen-doped carbon nanotubes [33]. This method involved a straightforward thermal treatment of ZIF-67 particles conducted at 700°C within a mixed Ar/H_2 atmosphere, followed by an acid treatment step to eliminate the embedded metal nanoparticles. Notably, the reduction-rich H_2 environment played a pivotal role in fostering the creation of intricate hierarchical carbon nanotube frameworks. The rapid formation of metallic Co nanoparticles acted as a catalyst for the growth of these nanotubes. Furthermore, beyond the synthesis of one-dimensional (1D) carbon nanotubes, notable advancements were made in the fabrication of two-dimensional (2D) carbon nanoribbons, as elucidated by Pachfule and colleagues in 2015 [34]. Initially, MOF-74 nanorods were meticulously prepared, followed by their transformation into carbon nanorods through controlled pyrolysis. Subsequently, these carbon nanorods underwent an exfoliation process, resulting in the creation of carbon nanoribbons. This novel type of carbon nanoribbon exhibited excellent electrochemical activity.

2.6 ACTIVATION

The activation step in the conversion of MOF precursors to carbon is a critical stage that significantly influences the final properties of the resultant carbon materials. This method involves the carbonization of MOFs followed by a subsequent activation step, often carried out with chemical agents or physical means. Activation helps to improve the porosity of the resulting carbon material by creating additional pores and increasing surface area. It plays a pivotal role in tailoring the structural, textural, and chemical properties of MOF-derived carbons. In short, activation methods hold the key to controlling the structure, porosity, and functionality of the resulting carbon. Activation techniques in the synthesis of MOF-derived carbon materials are essential for transforming MOF precursors into porous carbon structures with tailored properties. Here are some common activation techniques:

a. Chemical activation
    Chemical activation involves the use of chemical agents, such as strong acids (e.g., H_3PO_4) or alkalis (e.g., KOH), which are impregnated into MOF-derived materials. These chemicals react with the carbon to create additional pores and increase the surface area. Subsequent heating removes the chemical impurities, leaving behind a porous carbon structure.

b. Physical activation
    Physical activation methods include using steam or carbon dioxide as activating agents. In this process, MOF-derived materials are exposed to these gases at high temperatures. The gas activation creates pores and removes volatile components from the material, resulting in a porous carbon structure.

    Researchers often employ a combination of activation techniques to achieve specific properties. For example, a precursor may undergo both chemical and thermal activation to enhance pore development and surface area. The choice of activation method depends on the desired properties of the MOF-derived carbon material, including pore size, surface area, and functionality. Each activation technique offers a unique way to tailor the final carbon material to meet specific application requirements.

2.7 MOF-DERIVED METAL-BASED MATERIALS

MOFs have evolved as valuable precursors and templates for the production of metal nanoparticles/metal oxide/sulphide/phosphides, etc., along with metal/carbon composites, and metal oxide/carbon composites. MOFs offer a distinctive advantage as precursor materials for the synthesis of inorganic functional materials because of their ability to arrange metal cations in a precisely ordered spatial configuration, characterized by a meticulously defined chemical environment. In addition, MOF materials which serve as a source of porous carbon, upon in-situ carbonization, seamlessly combine with metal species. This synergistic approach elevates the pertinent properties of obtained materials, enhancing their performance in various applications. Furthermore, carbon when combined with inorganic oxides has emerged as a highly coveted functional material.

In an earlier study, Banerjee et al. presented the synthesis of a composite system, consisting of Fe_3O_4 nanoparticles embedded within a carbon matrix [35]. This composite was fabricated through the pyrolysis of an iron-containing MOF under an argon atmosphere. Specifically, the iron component, sourced from MIL-53, was subjected to annealing in an argon environment at two distinct temperatures, namely 500°C

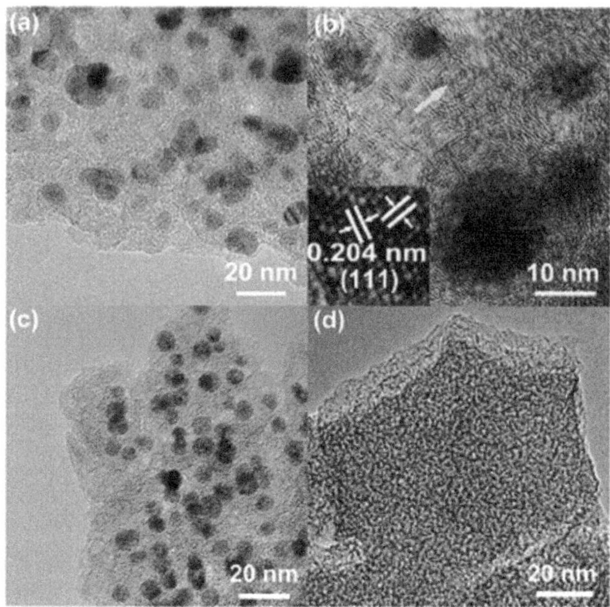

FIGURE 2.8 TEM images of MOF-derived catalysts: (a and b) ZIF-67-900, (c) $Co_2(bdc)_2$ (dabco)-900, (d) ZIF-8-900. Reproduced with permission from Ref. [37].

and 600°C. Remarkably, the resultant composite materials in each case exhibited the presence of rod-shaped particles, comprising Fe_3O_4 nanoparticles integrated within a carbon-rich matrix. Later, Ariga and co-workers developed for the first time, a hybrid of nanoporous carbon-cobalt oxide material by thermal conversion of ZIF-9 for electrocatalytic application [36] (Figure 2.8).

The preparation of metal nanoparticles by the direct thermal conversion of MOF is another important category in the MOF-derived metal-based materials. Wang et al. prepared MOF-derived surface-modified Ni nanoparticles by the pyrolysis of Ni-MOF in the presence of NH_3 in inert atmosphere [38]. In this study, the introduction of NH_4 during the pyrolysis stage plays a critical role in shaping the desired surface state conducive to hydrogen evolution reaction (HER). It achieves this by effectively eliminating excess carbon content and facilitating the essential surface nitridation process. The introduction of NH_3 results in metal nanoparticles without the presence of carbon matrix in the product. However, in order to prevent the agglomeration of metal nanoparticles which significantly decrease the activity of derived material, in most of the studies, these metal NPs are immobilized on solid supports such as carbon matrix, silica, and alumina. In this regard, carbon-encapsulated metal NPs synthesized by the MOF templating method holds a significant place. For example, Ahsan et al. prepared Cu and Ni NPs encapsulated in a carbon matrix by annealing Cu-BDC and Ni-BDC in Ar atmosphere at 600°C for 1 h [39]. The obtained magnetic C@Ni and C@Cu nanocatalysts exhibited a change in morphology compared to precursor MOF where the metal nanoparticles are distributed over the carbon sheet (Figure 2.9 and Table 2.4).

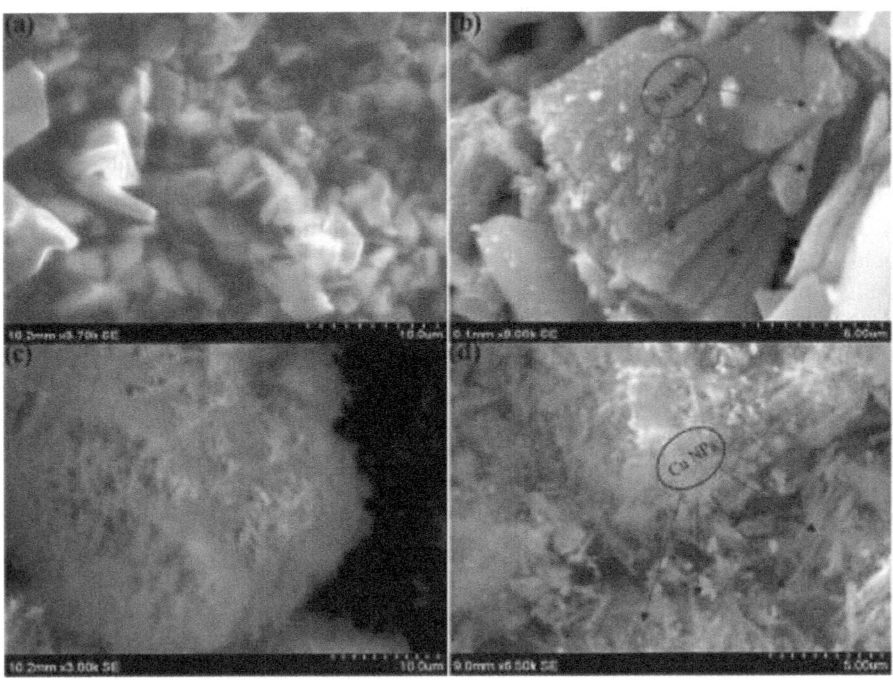

FIGURE 2.9 SEM micrographs of (a) Ni-BDC MOF; (b) C@Ni nanocatalyst; (c) Cu-BDC MOF; and (d) C@Cu nanocatalyst. Reproduced with permission from Ref. [39].

TABLE 2.4
Metal/metal oxide/sulphide/phosphide-decorated porous carbon prepared from MOF precursors

$M/MO_x/MS_x/MP_x$ nanostructure	Metal species	Dopants	Precursor MOF	Pyrolysis conditions	Ref.
$Co_2P/CoNPC$	Co_2 phosphide	Co, N, P	ZIF-67		[40]
Co/Co_9S_8@NSOC	Co/Co_9S_8 NPs	N, S, O	Co-NSO-MOF	600°C–900°C, N_2	[41]
Ni_3S_2@NPC	Ni_3S_2	N	Ni-MOF	400°C, Ar	[42]
Co_3O_4-NSP	Co_3O_4	N, S, Zn	Zn-DTO	–	[43]
$CoTe_2$@N-PC	$CoTe_2$	–	ZIF-67	500°C, 12 h, N_2	[44]
CoNi@NPC	CoNi NPs	–	CoNi-MOF	700°C, 2 h, N_2	[45]
Fe/N-PCNs	Fe	N	ZnFe-MOF	850°C–1050°C, 1 h, Ar	[46]
$MoSe_2$@MPCS	$MoSe_2$ nanoplates	–	Ni-MOF	800°C, 2 h, Ar	[47]
Fe_3O_4-C	Fe_3O_4 NPs	–	MIL-53	500°C–600°C, Ar	[35]

2.8 CONCLUSION

In summary, the resulting MOF-derived nanoporous carbons possess several advantageous features, including tunable pore sizes, well-defined structures, and controllable surface chemistry. These properties make them suitable for a wide range of applications. By carefully selecting the precursor MOF and optimizing the synthesis conditions, researchers can tailor the properties of the derived nanoporous carbons to meet specific application requirements. While MOF-derived carbon materials have achieved notable success, several persistent limitations emerge during the carbonization process: (1) the inherent predominance of micropores, (2) the irreversible aggregation of metal nanoparticles (NPs), and (3) the challenge of precisely controlling the structural changes. These factors collectively hinder their overall performance. In response to these technical constraints, a plethora of strategies are currently emerging to advance research in nanostructured carbon-based materials. These approaches encompass the development of mesoporous, hollow, yolk-shell, and multi-dimensional hollow or porous structures.

REFERENCES

1. Figueira, F.; Paz, F. A. A. Porphyrin MOF-Derived Porous Carbons: Preparation and Applications. *C*, 2021, *7* (2), 47.
2. Chen, L.; Bai, J.; Wang, C.; Pan, Y.; Scheer, M.; You, X. One-Step Solid-State Thermolysis of a Metal–Organic Framework: A Simple and Facile Route to Large-Scale of Multiwalled Carbon Nanotubes. *Chem Commun*, 2008, (13), 1581–1583.
3. (a) Yang, S. J.; Kim, T.; Im, J. H.; Kim, Y. S.; Lee, K.; Jung, H.; Park, C. R. MOF-Derived Hierarchically Porous Carbon with Exceptional Porosity and Hydrogen Storage Capacity. *Chem Mater*, 2012, *24* (3), 464–470; (b) Liu, Y.; Xu, X.; Shao, Z. Metal-Organic Frameworks Derived Porous Carbon, Metal Oxides and Metal Sulfides-Based Compounds for Supercapacitors Application. *Energy Storage Mater*, 2020, *26*, 1–22.
4. Radhakrishnan, L.; Reboul, J.; Furukawa, S.; Srinivasu, P.; Kitagawa, S.; Yamauchi, Y. Preparation of Microporous Carbon Fibers Through Carbonization of Al-Based Porous Coordination Polymer (Al-PCP) with Furfuryl Alcohol. *Chem Mater*, 2011, *23* (5), 1225–1231.
5. Zhong, S.; Zhan, C.; Cao, D. Zeolitic Imidazolate Framework-Derived Nitrogen-Doped Porous Carbons as High Performance Supercapacitor Electrode Materials. *Carbon*, 2015, *85*, 51–59.
6. Chen, Y.; Xie, Y.; Yan, X.; Cohen, M. L.; Zhang, S. Topological Carbon Materials: A New Perspective. *Phys Rep*, 2020, *868*, 1–32.
7. Liu, B.; Shioyama, H.; Akita, T.; Xu, Q. Metal-Organic Framework as a Template for Porous Carbon Synthesis. *J Am Chem Soc*, 2008, *130* (16), 5390–5391.
8. Liu, B.; Shioyama, H.; Jiang, H.; Zhang, X.; Xu, Q. Metal–Organic Framework (MOF) as a Template for Syntheses of Nanoporous Carbons as Electrode Materials for Supercapacitor. *Carbon*, 2010, *48* (2), 456–463.
9. Kyotani, T. Synthesis of Various Types of Nano Carbons Using the Template Technique. *Bull Chem Soc Jpn*, 2006, *79* (9), 1322–1337.
10. Guo, H.; Wang, M.; Zhao, L.; Youliwasi, N.; Liu, C. The Effect of Co and N of Porous Carbon-Based Materials Fabricated Via Sacrificial Templates MOFs on Improving

DA and UA Electrochemical Detection. *Microporous Mesoporous Mater*, 2018, *263*, 21–27.

11. Hu, J.; Wang, H.; Gao, Q.; Guo, H. Porous Carbons Prepared by Using Metal–Organic Framework as the Precursor for Supercapacitors. *Carbon*, 2010, *48* (12), 3599–3606.

12. Pachfule, P.; Biswal, B. P.; Banerjee, R. Control of Porosity by Using Isoreticular Zeolitic Imidazolate Frameworks (IRZIFs) as a Template for Porous Carbon Synthesis. *Chem Eur J*, 2012, *18* (36), 11399–11408.

13. Bai, F.; Xia, Y.; Chen, B.; Su, H.; Zhu, Y. Preparation and Carbon Dioxide Uptake Capacity of N-Doped Porous Carbon Materials Derived from Direct Carbonization of Zeolitic Imidazolate Framework. *Carbon*, 2014, *79*, 213–226.

14. Zou, G.; Jia, X.; Huang, Z.; Li, S.; Liao, H.; Hou, H.; Huang, L.; Ji, X. Cube-Shaped Porous Carbon Derived from MOF-5 as Advanced Material for Sodium-Ion Batteries. *Electrochim Acta*, 2016, *196*, 413–421.

15. Jeon, J.-W.; Sharma, R.; Meduri, P.; Arey, B. W.; Schaef, H. T.; Lutkenhaus, J. L.; Lemmon, J. P.; Thallapally, P. K.; Nandasiri, M. I.; McGrail, B. P. In Situ One-Step Synthesis of Hierarchical Nitrogen-Doped Porous Carbon for High-Performance Supercapacitors. *ACS Appl Mater Interfaces*, 2014, *6* (10), 7214–7222.

16. Hao, L.; Wei, J.; Zheng, R.; Wang, C.; Wu, Q.; Wang, Z. Magnetic Porous Carbon Derived from Co-Doped Metal–Organic Frameworks for the Magnetic Solid-Phase Extraction of Endocrine Disrupting Chemicals. *J Sep Sci*, 2017, *40* (20), 3969–3975.

17. Carrasco, J.; Romero, J.; Abellán, G.; Hernández-Saz, J.; Molina, S.; Martí-Gastaldo, C.; Coronado, E. Small-Pore Driven High Capacitance in a Hierarchical Carbon Via Carbonization of Ni-MOF-74 at Low Temperatures. *Chem Commun*, 2016, *52* (58), 9141–9144.

18. Zhong, H. X.; Wang, J.; Zhang, Y. W.; Xu, W. L.; Xing, W.; Xu, D.; Zhang, Y. F.; Zhang, X. B. ZIF-8 Derived Graphene-Based Nitrogen-Doped Porous Carbon Sheets as Highly Efficient and Durable Oxygen Reduction Electrocatalysts. *Angew Chem Int Ed*, 2014, *53* (51), 14235–14239.

19. Liu, S.; Zhang, H.; Zhao, Q.; Zhang, X.; Liu, R.; Ge, X.; Wang, G.; Zhao, H.; Cai, W. Metal-Organic Framework Derived Nitrogen-Doped Porous Carbon@ Graphene Sandwich-Like Structured Composites as Bifunctional Electrocatalysts for Oxygen Reduction and Evolution Reactions. *Carbon*, 2016, *106*, 74–83.

20. Zhang, X.; Luo, J.; Tang, P.; Morante, J. R.; Arbiol, J.; Xu, C.; Li, Q.; Fransaer, J. Ultrasensitive Binder-Free Glucose Sensors Based on the Pyrolysis of In Situ Grown Cu MOF. *Sens Actuators, B*, 2018, *254*, 272–281.

21. Zhang, Z.; Huo, H.; Gao, J.; Yu, Z.; Ran, F.; Guo, L.; Lou, S.; Mu, T.; Yin, X.; Wang, Q. Ni-MOF Derived NiO/C Nanospheres Grown In Situ on Reduced Graphene Oxide Towards High Performance Hybrid Supercapacitor. *J Alloys Compd*, 2019, *801*, 158–165.

22. Li, Z.; Shao, M.; Zhou, L.; Zhang, R.; Zhang, C.; Wei, M.; Evans, D. G.; Duan, X. Directed Growth of Metal-Organic Frameworks and Their Derived Carbon-Based Network for Efficient Electrocatalytic Oxygen Reduction. *Adv Mater*, 2016, *28* (12), 2337–2344.

23. Su, J.; Yang, Y.; Xia, G.; Chen, J.; Jiang, P.; Chen, Q. Ruthenium-Cobalt Nanoalloys Encapsulated in Nitrogen-Doped Graphene as Active Electrocatalysts for Producing Hydrogen in Alkaline Media. *Nat Commun*, 2017, *8* (1), 14969.

24. Zhang, C.; Wang, Y. C.; An, B.; Huang, R.; Wang, C.; Zhou, Z.; Lin, W. Networking Pyrolyzed Zeolitic Imidazolate Frameworks by Carbon Nanotubes Improves

Conductivity and Enhances Oxygen-Reduction Performance in Polymer-Electrolyte-Membrane Fuel Cells. *Adv Mater*, 2017, *29* (4), 1604556.

25. Gadipelli, S.; Li, Z.; Zhao, T.; Yang, Y.; Yildirim, T.; Guo, Z. Graphitic Nanostructures in a Porous Carbon Framework Significantly Enhance Electrocatalytic Oxygen Evolution. *J Mater Chem A*, 2017, *5* (47), 24686–24694.

26. Yang, L.; Zeng, X.; Wang, W.; Cao, D. Recent Progress in MOF-Derived, Heteroatom-Doped Porous Carbons as Highly Efficient Electrocatalysts for Oxygen Reduction Reaction in Fuel Cells. *Adv Funct Mater*, 2018, *28* (7), 1704537.

27. Chen, B.; Yang, Z.; Ma, G.; Kong, D.; Xiong, W.; Wang, J.; Zhu, Y.; Xia, Y. Heteroatom-Doped Porous Carbons with Enhanced Carbon Dioxide Uptake and Excellent Methylene Blue Adsorption Capacities. *Microporous Mesoporous Mater*, 2018, *257*, 1–8.

28. Ye, C.; Xu, L. Heteroatom-Doped Porous Carbon Derived from Zeolite Imidazole Framework/Polymer Core-Shell Fibers as an Electrode Material for Supercapacitor. *Composites, Part B*, 2021, *225*, 109256.

29. Van Nguyen, C.; Lee, S.; Chung, Y. G.; Chiang, W.-H.; Wu, K. C.-W., Synergistic Effect of Metal-Organic Framework-Derived Boron and Nitrogen Heteroatom-Doped Three-Dimensional Porous Carbons for Precious-Metal-Free Catalytic Reduction of Nitroarenes. *Appl Catal, B*, 2019, *257*, 117888.

30. Liu, W.; Li, S.-Q.; Liu, W.-X.; Zhang, Q.; Shao, J.; Tian, J.-L. MOF-Derived B, N Co-Doped Porous Carbons as Metal-Free Catalysts for Highly Efficient Nitro Aromatics Reduction. *J Environ Chem Eng*, 2021, *9* (4), 105689.

31. Xu, D.; Ding, Q.; Li, J.; Chen, H.; Pan, Y.; Liu, J. A Sheet-Like MOF-Derived Phosphorus-Doped Porous Carbons for Supercapacitor Electrode Materials. *Inorg Chem Commun*, 2020, *119*, 108141.

32. Dilpazir, S.; Liu, R.; Yuan, M.; Imran, M.; Liu, Z.; Xie, Y.; Zhao, H.; Zhang, G. Br/Co/N Co-Doped Porous Carbon Frameworks with Enriched Defects for High-Performance Electrocatalysis. *J Mater Chem A*, 2020, *8* (21), 10865–10874.

33. Xia, B. Y.; Yan, Y.; Li, N.; Wu, H. B.; Lou, X. W. D.; Wang, X. A Metal–Organic Framework-Derived Bifunctional Oxygen Electrocatalyst. *Nat Energy*, 2016, *1* (1), 1–8.

34. Pachfule, P.; Shinde, D.; Majumder, M.; Xu, Q. Fabrication of Carbon Nanorods and Graphene Nanoribbons from a Metal–Organic Framework. *Nat Chem*, 2016, *8* (7), 718–724.

35. Banerjee, A.; Gokhale, R.; Bhatnagar, S.; Jog, J.; Bhardwaj, M.; Lefez, B.; Hannoyer, B.; Ogale, S. MOF Derived Porous Carbon–Fe3O4 Nanocomposite as a High Performance, Recyclable Environmental Superadsorbent. *J Mater Chem*, 2012, *22* (37), 19694–19699.

36. Chaikittisilp, W.; Torad, N. L.; Li, C.; Imura, M.; Suzuki, N.; Ishihara, S.; Ariga, K.; Yamauchi, Y. Synthesis of Nanoporous Carbon–Cobalt-Oxide Hybrid Electrocatalysts by Thermal Conversion of Metal–Organic Frameworks. *Chem Eur J*, 2014, *20* (15), 4217–4221.

37. Wang, X.; Zhou, J.; Fu, H.; Li, W.; Fan, X.; Xin, G.; Zheng, J.; Li, X. MOF Derived Catalysts for Electrochemical Oxygen Reduction. *J Mater Chem A*, 2014, *2* (34), 14064–14070.

38. Wang, T.; Zhou, Q.; Wang, X.; Zheng, J.; Li, X. MOF-Derived Surface Modified Ni Nanoparticles as an Efficient Catalyst for the Hydrogen Evolution Reaction. *J Mater Chem A*, 2015, *3* (32), 16435–16439.

39. Ahsan, M. A.; Jabbari, V.; El-Gendy, A. A.; Curry, M. L.; Noveron, J. C. Ultrafast Catalytic Reduction of Environmental Pollutants in Water Via MOF-Derived Magnetic

Ni and Cu Nanoparticles Encapsulated in Porous Carbon. *Appl Surf Sci*, 2019, *497*, 143608.

40. Liu, H.; Guan, J.; Yang, S.; Yu, Y.; Shao, R.; Zhang, Z.; Dou, M.; Wang, F.; Xu, Q. Metal–Organic-Framework-Derived Co2P Nanoparticle/Multi-Doped Porous Carbon as a Trifunctional Electrocatalyst. *Adv Mater*, 2020, *32* (36), 2003649.

41. Du, J.; Wang, R.; Lv, Y.-R.; Wei, Y.-L.; Zang, S.-Q. One-Step MOF-Derived Co/Co9S8 Nanoparticles Embedded in Nitrogen, Sulfur and Oxygen Ternary-Doped Porous Carbon: An Efficient Electrocatalyst for Overall Water Splitting. *Chem Commun*, 2019, *55* (22), 3203–3206.

42. Yang, M.; Ning, Q.; Fan, C.; Wu, X. Large-Scale Ni-MOF Derived Ni3S2 Nanocrystals Embedded in N-Doped Porous Carbon Nanoparticles for High-Rate Na+ Storage. *Chin Chem Lett*, 2021, *32* (2), 895–899.

43. Pal, S.; Jana, S.; Kumar, A.; Prakash, R. Enhanced OER Properties from Nanocomposites of Co3O4 and MOF Derived N/S/Zn-Doped Porous Carbon. *Electrochim Acta*, 2022, *436*, 141436.

44. Zhang, B.; Zhang, Y.; Li, J.; Liu, J.; Huo, X.; Kang, F. In Situ Growth of Metal–Organic Framework-Derived CoTe 2 Nanoparticles@ Nitrogen-Doped Porous Carbon Polyhedral Composites as Novel Cathodes for Rechargeable Aluminum-Ion Batteries. *J Mater Chem A*, 2020, *8* (11), 5535–5545.

45. Li, J.; Xue, H.; Xu, N.; Zhang, X.; Wang, Y.; He, R.; Huang, H.; Qiao, J. Co/Ni Dual-Metal Embedded in Heteroatom Doped Porous Carbon Core-Shell Bifunctional Electrocatalyst for Rechargeable Zn-Air Batteries. *Mater Rep: Energy*, 2022, *2* (2), 100090.

46. Zheng, L.; Yu, S.; Lu, X.; Fan, W.; Chi, B.; Ye, Y.; Shi, X.; Zeng, J.; Li, X.; Liao, S. Two-Dimensional Bimetallic Zn/Fe-Metal-Organic Framework (MOF)-Derived Porous Carbon Nanosheets with a High Density of Single/Paired Fe Atoms as High-Performance Oxygen Reduction Catalysts. *ACS Appl Mater Interfaces*, 2020, *12* (12), 13878–13887.

47. Zhao, W.; Ma, X. Hierarchical Scalelike Yolk–Shell Construction Assembled Via Ultrathin MoSe2 Nanoplates Incorporated into Metal–Organic Frameworks Derived Porous Carbon Spheres as Highly Durable Anode for Enhanced Sodium Storage. *ACS Sustainable Chem Eng*, 2020, *8* (51), 19040–19050.

3 Metal–Organic Framework Derived Materials
Structure and Properties

3.1 INTRODUCTION

The structure-property relationship is a fundamental principle in the realm of materials science and engineering. It signifies the connection between a material's atomic or molecular structure and its physical, chemical, and mechanical properties. Understanding this connection is crucial for creating and customizing materials with specific attributes tailored for various practical purposes. Understanding how factors like the arrangement of atoms or molecules influence a material's properties empowers scientists and engineers to develop advanced materials that satisfy a wide range of application requirements such as adsorption, gas storage, catalysis, energy storage, and sensing. A schematic depicting the relationship among structure, properties, processing, and performance is shown in Figure 3.1.

MOFs are available in diverse architectures and features. Distinct functionalities can be obtained by tailoring the substituents that make up the framework. Hence, they can be adapted as precursors or templates to obtain a wide variety of nanoarchitectures. Several post-modification techniques can be employed to derive nanostructures from these MOFs. The derived materials possess high surface areas and thermal and chemical stabilities and exhibit enhanced performance than the pristine MOFs. As discussed in earlier chapters, several techniques are available to obtain derived structures. Among the methods contributing to the formation of MOF-derived structures, a direct thermal transformation is a simple strategy to convert MOFs into nanostructures. The main strategies involved in converting the MOFs are templating strategy and the morphology retention strategy. Again, the templating method can be further divided into self-templating and external templating (use of external agents such as silica, graphene, and metal oxides). By careful selection of precursors and design strategies, a wide range of structures with diversified morphologies can be obtained, such as zero-dimensional core-shell, hollow, and polyhedral; one-dimensional rods and tubes; two-dimensional sheets and nanoribbons; and three-dimensional cages, honeycomb structures, arrays, etc. The morphology of the resulting MOF-derived materials depends upon the morphology of the template used.

DOI: 10.1201/9781003432357-4

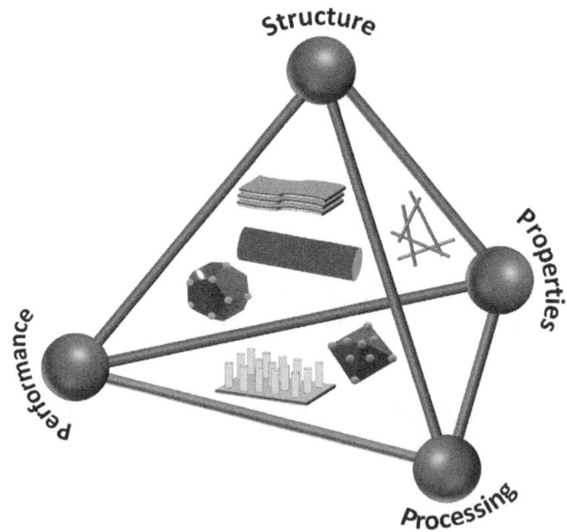

FIGURE 3.1 Schematic illustration of structure, properties, processing, and performance relationship.

It is vital to understand how a material's morphology affects its characteristics and behavior, to improve the performance for specific applications. The morphology of nanoparticles or nanoscale objects is important in obtaining specific characteristics for targeted applications. The reactivity, catalytic activity, and biological interactions of these materials may all be affected by their size, shape, and surface features. For example, in the case of CO_2 adsorption, regulating the morphology and structure will improve the active sites for enhanced CO_2 adsorption and affinity between the adsorbate and adsorbent. In the case of catalysis, the main aim of post-synthetic modification of MOF precursor is to obtain definitive derived structures for better catalytic activity by reducing Tafel slopes, overpotential, and increasing the current densities. Besides, precise regulation of morphology will enhance surface area and pore size distribution, leading to efficient electron and mass transport. Due to this, the performance of the material will be phenomenally improved, especially in applications where charge transport is significant. Apart from that, it is necessary to enhance the diffusion rates by achieving shortest paths for diffusion of ions, guest species, and any other key performance contributors. To achieve these kind of results, one of the strategy like heteroatom doping was proposed to enable host-guest interactions, alter the electronic properties, and thus improve the performance. During this process, it is important to preserve the structure without compromising on these required functionalities. Practically, it is very difficult to achieve this. Hence, regulating the morphology will be an alternative approach to accomplish the definite electronic, physical, and chemical properties in MOF-derived materials.

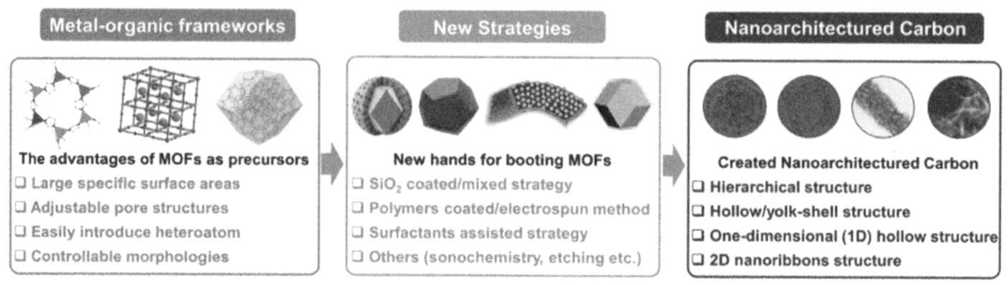

FIGURE 3.2 Advantages of MOFs as precursors and emerging strategies involved in engineering the MOF-derived materials. Reproduced with permission from Ref. [1].

3.2 FACTORS AFFECTING THE STRUCTURE AND MORPHOLOGY

There are several factors affecting the structure and morphology, thereby properties and performance of MOF-derived materials. These include the following.

1. Precise selection of MOF precursor with definite metal coordination centre
2. Proper creation of ligand environment
3. Synthesis conditions such as temperature, time, and gas atmosphere
4. Synthesis route
5. Incorporation of additional agents that assist in developing new functional materials

Precise regulation of these factors will impact the physiochemical properties leading to multiple nanoarchitectures: hollow structures, nanoribbons, nanospheres, nanocages, yolk/shell structures, hierarchical porous structures, and many more. The following sections will elaborate on different morphologies of MOF-derived materials, various factors influencing them, and strategies involved in engineering the structure and morphology of these materials (Figure 3.2).

3.3 VARIOUS MORPHOLOGIES OF MOF-DERIVED MATERIALS

MOFs with nanoscale dimensions can be created using controlled strategies to obtain well-defined structures as shown in Figure 3.3. Upon various treatments, these MOFs can be transformed into the derived architectures while preserving the morphology. For this purpose, the MOFs should be thermally stable enough at least up to the temperatures above which they transform. Otherwise, their structures may collapse at temperatures which are well below the conversion range [2]. The following sections will provide a brief description of various morphologies governing MOF-derived materials.

3.3.1 ZERO-DIMENSIONAL ARCHITECTURES

Zero-dimensional structures are those materials with all the dimensions in the nano-scale range. Typical examples include carbon dots, N-GQDs, metal NPs, etc. which

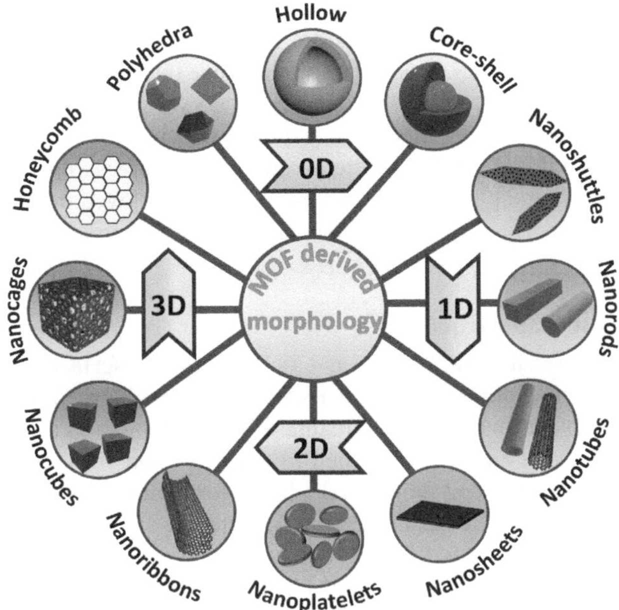

FIGURE 3.3 Schematic illustration of various morphologies of MOF-derived materials.

exist in different forms such as oxides, phosphides, halides, and many more. These structures exist in various morphologies which can be mainly divided into three types: polyhedra, hollow, and core-shell architectures. The rational design of these materials with addition of hetero-doping of atoms paves a way in improving the textural, chemical, and electronic properties.

3.3.1.1 Polyhedra

Polyhedra are among the most typical forms of MOF-derived nanostructures such as tetrahedral, octahedral, and dodecahedral. Notable structures like ZIFs [3], Zr (CPM-99) [4], and Mb-MOF [5] serve as a template to obtain polyhedral architectures by well retaining the original morphology with huge number of edges and corners which contribute for the formation of active sites. Heteroatom-doped polyhedral structures were used in various applications such as H_2 production, oxygen evolution reaction (OER), oxygen reduction reaction (ORR), and batteries. Selection of suitable MOF precursor will be a deciding factor in obtaining preferred polyhedral morphologies. ZIF-8 and ZIF-67 showed promising results to give single atom dispersed porous carbon polyhedra. However, there were several challenges encountered to materialize the ZIF into porous carbon with highly dispersed active sites and high concentrations. One of the solution provided was to integrate ZIF-8 and ZIF-67 into a bimetallic Zn-Co-MOF to give a porous carbon upon pyrolysis with high surface area, high N-content of ZIF-8, and greater graphitization degree of ZIF-67 [6]. Apart from the ZIF morphologies, other categories such as micro-boxes (cubical) and octahedral were also reported using different MOF precursors based on Zn, Co, Cu, Fe, and many more. For these MOF precursors, simple and conventional strategies can be utilized to obtain the desired polyhedra structures. However, for some kinds of MOFs such as Mo, W, and V, it is difficult to achieve the intended morphology after the derivation process. Introduction of guest species will enable confined carburization process and aid in the synthesis of required structures [5]. Recently, N-doped graphene quantum dots (N-GQD) were derived from ZIF-8 using a pioneering acid vapor cutting technique [7]. A similar technique was employed in the preparation of carbon nano-polyhedrons for fluorescence applications such as detection of heavy metal ions.

3.3.1.2 Hollow Nanostructures

Hollow architectures are low-density materials with huge surface areas and can be loaded with high concentrations of guest species. External templating strategies are utilized in the preparation of hollow structures. Suitable templates are selected, coated with MOF layers, and are then subjected to pyrolysis and subsequent etching process. The templates act as sacrificial media in the formation of the hollow structures. The textural features of the resulting hollow material, size, and shape of the void will greatly depend upon the selected template and the thickness of MOF coating. Several mechanisms such as Kirkendall effect, Ostwald ripening, thermal decomposition, chemical etching, and galvanic replacement were used to prepare hollow structures. The hollow nature to the resulting material was contributed mainly by volume shrinkage and the release of gases like NO_x, CO_2, and H_2O during the heat treatment process. The formation of hollow framework is usually governed by a non-equilibrium interdiffusion process. The attainment of a stable structure greatly depends upon the diffusion rates during thermal decomposition. If the diffusion rates and the volume of gases released are too high, the hollow structure gets collapsed. Hollow architectures, in turn, exist in several shapes such as parallelopipeds, micro boxes, hexagonal, ellipsoidal, rod-shaped, and many more. The formation of these shapes depends on the initial precursor chosen and the parameters set forth. Apart from external templating, a self-templating strategy was mostly utilized in the preparation of hollow structures using MOF alone as the substrate material. This reduces the overall cost and makes the process simple. Hollow structures find a wide range of opportunities, especially in manufacture of lithium-ion batteries due to the easier accessibility of the Li^+ ions with shorter diffusion paths and also helps in relieving the stresses during continuous battery operation.

3.3.1.3 Core-Shell Nanostructures

Core-shell materials are composed of an inner core and outer shell. These type of structures are usually formed from pre-existing core-shell MOFs. The resulting material exhibits good performance and stability. The core and the shell components are structurally so robust that there always exists a synergistic effect between them which makes the material show potential functionalities. In most cases, the inner core aids in the functionality and performance of the material while the outer shell acts as a permeable membrane and also protects the inner core from deactivation. Thus, structural robustness is preserved in these structures. Multi-shell structures have also been prepared especially for the electrocatalytic applications to enhance the structural robustness of the catalyst. Due to the presence of multi-shells, the catalyst can accommodate the changes in the volume of pores with improved stability. Similar to single-shell structures, the outer shells of these structures prevent the catalyst from deactivation whereas the inner shell promotes the catalytic activity.

3.3.2 ONE-DIMENSIONAL ARCHITECTURES

One-dimensional nano structures possess at least one of its dimension in the nanoscale range. Typical examples include nanotubes, nanorods, nano shuttles, nanowires, etc. Wide variety of these structures can be obtained in the form of oxides, sulphides, halides, composites, etc. The physical properties of these structures gained utmost interest which made them use in emerging applications such as opto-electronics and drug delivery. They can be prepared using a templated method or without the use of template. Hierarchical porous composite 1D nano structures were also prepared using specific strategies. In the below sections, some of the known 1D architectures have been discussed which showed promising results in potential applications.

3.3.2.1 Nanorods

1D nanorods can be prepared by use of the MOF precursors alone without employing any template. Both mono and bimetallic MOF precursors belong to MOF-74, ZIF families were used to prepared nanorods. Mostly, they were synthesized using the MOFs with Al [8, 9], Co [10, 11], Fe [12], Zn [13] and Ni [14, 15] as the metal centres. These metal-based MOFs exist in rod like morphology which can be directly used as a precursor in the preparation of nanorods. The structure of MOF-74 contains hexagonal channels which aid in the formation of 1D structures. Pyrolysis of these MOFs and subsequent post-synthetic treatments results in the formation of nanorods. Acid etching of these nanorods results in the formation of nanoribbons under controlled conditions. Nanorods are specifically used in energy applications such as batteries, supercapacitors, drug delivery and sensing applications. These structures facilitate efficient ion transport, charge transport with lesser distances and shortest diffusion times.

3.3.2.2 Nanotubes

Nanotubes, one of the emerging class of nanomaterials, have been applied in wide variety of applications. They exhibit very high surface area, stability, strength and has good transport properties. They act as fillers in the preparation of composite materials which improves their adhesion properties, mechanical strength, conductivity, and in polymeric membrane materials which improves their permeability, separation ability and transport characteristics. SWCNT, MWCNT, TiO_2 tubes, and many other tubular species were explored in applications such as catalysis, sensing, gas separation, adsorption, and energy storage. MOFs act as self-template in the preparation of nanotubes. Additionally, template-strategy can also be used to produce nanotubes. Similar to the synthesis strategy of nanorods, nanotubes can be produced from Fe,

Co, Ni-based MOFs due to their tubular nature. Generally chemical vapour deposition is used for preparing CNTs which is an expensive and time-consuming process. Alternatively, MOF templated synthesis will make the process simple and viable. However, there are challenges in the self-templating strategy such as controlling the dimensions of CNTs, uniformity in the distribution of doped species and the degree of graphitization. Careful synthetic procedures and suitable MOF precursors help in achieving good results.

3.3.2.3 Nano Shuttles

Shuttle-like morphologies were observed during the pyrolysis of several MOFs. They contain large specific surface area and can accommodate huge amount of active centres distributed uniformly throughout. Most prominent shuttle-like structure is the hexagonal spindle morphology with porous nature. Possessing these characteristics, nano-shuttle like morphologies were found to be applied in various applications such as lithium-ion batteries [16], sodium ion batteries [17], and capacitive deionization [18]. Due to their unique morphology, it exhibits excellent current densities, low charge transfer resistances and long-term cyclic stability.

3.3.3 Two-Dimensional Architectures

Two-dimensional structures are composed of thin layers with thickness in the range of nanometres. They possess high surface areas and contain abundant active sites that can participate in several reactions. They can be applied in applications such as separation, adsorption, optoelectronic devices, and energy storage devices. Typical examples include nanoplatelets, nanosheets, nanoflakes, nanoribbons, etc. They can be synthesized using either self-templating or external templating strategies. In most cases, lamellar shaped MOF materials were preferred as precursors in synthesizing 2D structures. Other procedures were also explored which employ thermal transformation followed by post-treatment methods. However, direct conversion of MOF precursors into 2D structures is a feasible route instead of utilizing post-treatment methods. To produce high quality 2D materials, external templates can be coupled with MOFs facilitating the uniform growth of the resulting material.

3.3.3.1 Nanosheets

A wide variety of nanosheets such as TiO_2 nanosheets and graphene nanosheets can be synthesized using MOFs as the precursors. Two-dimensional structures derived from MOFs possess high surface area and porosity, rich active sites, high conductivity, and low charge transfer resistance. During the thermal transformation process, due to the collapse of the channels, there is a chance of metal aggregation. However, this aggregation is least when compared with that of 3D structures [19]. 2D nanosheets had been applied in several catalytic applications such as OER, HER, and CO_2 reduction. Wide variety of structures such as metal oxide/hydroxide nanosheets, metal sulphide and metal phosphide nanosheets, and metal boride nanosheets. The type of nanosheet obtained depends upon the MOF precursor used, corresponding source for S, P or B employed during thermal transformation and the pyrolysis conditions.

3.3.3.2 Nanoplatelets

Two-dimensional nanoplatelets can be synthesized under controlled conditions. They exhibit layered morphology with considerable thickness with lateral dimensions in the micrometre range. Due to the difficulties in controlling the pyrolysis conditions, only a few studies were reported on these kind of morphologies. One of them is the application of Sn-MOF@Se decorated with N doped graphene for sodium ion storage [20]. In this, MOF served as a template to synthesize SnSe/C structure which was efficient and provided an alternate path for the traditional synthesis route of SnSe structures. Due to the versatile inner and outer carbon shell frameworks, the stability was improved. Other instance was the synthesis of cuprous oxide nanoplatelets for dye degradation from Cu-MOF [21]. Here controlled morphology was obtained using PVP as the inhibitor which directed the Cu species to coordinate with the bipyridine ligand. Due to the addition of PVP, the stacking of layers was prevented.

3.3.3.3 Nanoribbons

Nanoribbons, especially graphene nanoribbons, are the strips of graphene with dimensions in the nanometre range. They are sometimes classified as 1D structures, but mostly reported under 2D nanostructures. These are formed by the chemical exfoliation of porous carbon rods obtained from pyrolysis of MOF precursors. Initially rod-based structures can be formed from MOF precursors such as MOF-74. After pyrolysis, the porous carbon rods are exfoliated to give layered graphene nanoribbons.

3.3.4 THREE-DIMENSIONAL STRUCTURES

Nano materials occurring in certain shapes such as nano cubes, nano sponges, nanocages, and honeycomb structures are termed as three-dimensional (3D) nano architectures. They exist as nano crystals in bulk materials. When compared to 0D, 1D and 2D materials exhibit higher specific surface areas, excellent interactions with guest species and possess interconnectivity inside the framework. The following sections will explore various 3D morphologies attained so far.

3.3.4.1 Nano Cubes

Cubic shaped porous carbon networks or composites were prepared at early stages by using MOF-5 as the precursor. This Zn-based MOF retained its original cubic morphology by transforming into ZnO/C nanocomposite [22]. Due to excellent π-π interactions, the material showed profound photocatalytic degradation of pollutants. Nano cubes can be prepared using other MOFs as precursors such as ZIF-8, UiO-66 and ZIF-8 using modified solvothermal synthesis. Other approaches such as wet chemical methods can also be used to prepare 3D nano cubes by regulating the concentrations of starting materials and pH metrics.

3.3.4.2 Nanocages and Nano Sponges

Nano sponges and nano cages can exhibit well accessible active sites that can be applied in catalysis. They can be derived from rationally designed templates and coupling with MOF precursors. Sometimes, these derived materials may exhibit low dispersion in certain solvents. Hence, these composites can be grown on some external agents such as carbon sponges. For example, ZnO nano cages were grown on rGO coated carbon sponge [23]. ZIF-8 was grown on 3D graphene network [24]. The pyrolysis of these materials resulted in formation of 3D nanocomposites that showed hierarchical porosity and well-equipped reactive centres.

3.3.4.3 Honeycomb Structures

Honeycomb structures are unique class of morphologies that can be obtained by rationally designed strategies. In a typical synthesis reported, ZIF-67 was grown on Co-Al layered double hydroxide (LDH) and subsequently pyrolyzed to give honeycomb like porous carbon network [25]. Instead of Co-Al-LDH, other supports like $Co(OH)_2$ and AlOOH were also explored but could not result in porous honeycomb structure. Even ZIF-8 was tried instead of ZIF-67 as a precursor which was a failure. This suggests that the choice of specific template and precursor is necessary to direct the uniform growth of the nanocrystals. Other supports were also reported exclusively which gave porous honeycomb like structures.

3.4 EFFECTS OF PROCESSING PARAMETERS ON THE STRUCTURE AND PROPERTIES OF MOF-DERIVED MATERIALS

Processing parameters have a significant impact on the structure and properties of the resulting MOF-derived materials. The precise regulation of these parameters will contribute in optimizing the characteristic features of the material such as specific surface area, pore size, conductivity, charge, and mass transport and many other properties. A brief illustration of the parameters affecting the outcome MOF-derived materials is explained below.

3.4.1 Effect of Carbonization Temperature

Temperature has a significant contribution in controlling the structure, morphology, and composition of the derived material. It plays a dominant role in regulating the particle size, surface area, percentage of heteroatom doping and N content. Hetero atom doping such as N, P and S act as highly active sites especially in applications like adsorption, catalysis, and sensing. Increasing the temperature leads to reduction in the percentage of heteroatom doping. Temperature has an inevitable impact on the source and amount of N content present in the sample after carbonization. The general source of N content such as pyridinic N, pyrrolic N, and graphitic N imparts specific functionalities to the material. The former two take part in the coordination with the metal atoms while the latter contributes to the electronic properties of the material. With increasing carbonization temperature, pyrrolic N and pyridinic N transforms into oxidized N and graphitic N. Besides these effects, temperature can also control the morphology of the material leading to formations of nanorods, nanospheres, core-shell, hierarchical and various other structures. It can also regulate the porosity and specific surface area (SSA) of the material facilitating effective mass transport. Nevertheless, high carbonization temperatures can sometimes lead to collapse of the structure. Hence careful selection of carbonization temperature will significantly impact the structure, morphology, and composition of the MOF-derived material.

When the MOF is subjected to heat treatment, the metal ions become highly energetic and mobile within the framework. Higher temperature results in faster diffusion of metal ions and can cause agglomeration. Temperature must be adjusted such that significant metal diffusion takes place without any aggregation. This temperature varies with different metals. There are two parameters defined to get an appreciable knowledge on controlling the temperature for precise control of the structure. One is the Tammann temperature which is equal to half of the melting point of metal centres in Kelvin scale. At this temperature, the bulk diffusion of the metal species will be significant. The other parameter is the Huttig temperature which is equal to one-third of the melting point of metal centres in Kelvin scale. It is the temperature at which the agglomeration becomes significant. Hence, careful selection of temperature must be done to ensure the formation of desired structure [26].

In one of the works, Fe incorporated Co-MOFs were pyrolyzed at various temperatures starting from 800°C to 1000°C [27]. The initial cage morphology Co-MOF was transformed into N-doped Fe_3C carbon networks with various morphologies such as tubular carbon, onion like carbon and tubular graphene/N-doped graphene. In another work, melamine incorporated Fe-MIL-88 was pyrolyzed at different temperatures [28]. As the temperature started increasing, the layer like morphology began to curl and transform into nanotubes. This is due to the formation of Fe_3C and its growth confinement as the temperature is increased. Thus, it can be concluded that temperature has a vital impact on structure and morphology.

3.4.2 EFFECT OF GASEOUS ENVIRONMENT

The composition and the structure of the MOF-derived material is highly effected by the type of gaseous environment present during the heat treatment process. Generally, the pyrolysis takes place in a one-step procedure. However, in some cases, the pyrolysis can take place in two steps or multiple steps employing different gaseous environments in each of the steps. Various gases employed during the heat treatment are Air, O_2, inert gases such as N_2 and Ar. Other gases such as C_2H_2, NH_3, and H_2 can also be employed to regulate the composition and structure of MOF-derived material. In an inert atmosphere like N_2, the heat treatment results in the removal of O, H from the organic linker in the form of water molecules leading to dense carbon structure. Other elements like N, P, S if exist in the ligand, will remain as the dopants. In the presence of O_2 rich atmospheres such as air or O_2, the carbon present in the organic ligand readily undergoes oxidation and may escape as CO_2/CO. The structure will be composed of metal oxide nanoparticles. Metal carbide formation may take place in the presence of reducing gas atmospheres. Similarly, metal sulphides and metal phosphides can be obtained during heat treatment in phosphorous and sulphur containing environments, respectively. The use of NH_3 results in the formation of metal nitrides in the final structure. Hence it can be understood that the gaseous environment not only acts as heat transferring agent but also takes part in chemical reactions. However, this is not true for all kinds of MOFs. It was observed that the type of structure formed was dependent on the standard reduction potential of metal species. The metals with high reduction potential were less prone to oxidation while those with low reduction potential were prone to oxidation to form metal oxide structures [29].

Mixed gas atmospheres strategy was employed to engineer the morphology of the MOF-derived materials. A Co-Zn-MOF was pyrolyzed in the presence of C_2H_2/ N_2 mixed atmosphere [30]. Interestingly it was observed that the formation of CNT occurred in the mixed atmosphere whereas no CNT formation took place in N_2 alone atmosphere. C_2H_2 contributed as a source of carbon for CNT growth. Hence it can be concluded that gaseous environment plays a crucial role in altering the morphology of MOF-derived materials.

3.4.3 Effect of Heating Rate

Heating rate has a huge impact on the structure and the characteristic features of the MOF-derived materials. Heating rate affects the extent of graphitization of organic parts and also the particle size of the final structure. Low heating rates like 1 and 2°C/min results in higher porosity and less particle size. The lower heating rates will increase the heat treatment time and leads to evaporation of unstable components. Higher heating rates such as 10°C/min results in agglomeration of the particles. In one of the studies, it was observed that the heating rate significantly impacted the morphology of the derived material. Chen and his co-workers synthesized a MOF-derived material from a Zn/Ni bimetallic MOF to study the electrochemical activity [31]. Heating rate of about 1°C/min resulted in the formation of ZnO/NiO microsphere like nanoparticles, while heating rate of 5°C/min produced a seaweed like nanosheet morphology of ZnO/NiO. Since a time was available at low heating rate, it led to sintering process giving rise to the decomposition of precursors and growth of nanoparticles. The sheet like morphology contributed to the high electronic and ionic conductivity due to which the best electrochemical performance was achieved with a capacitance of 435.1 F/g. The effect of heating rate on the oxygen reduction reaction (ORR) studied by Cui et al. had shown interesting results which can be closely related to the performance of the catalyst [32]. The material was subjected to different heating rates of 10, 5, and 2°C/min. The sample with a slow heating rate showed outstanding performance with a half-wave potential of 0.80 VRHE and a current density of 5 mA/cm^2. From the SEM micrographs, it can be understood that, with the increase of heating rate, the structure was composed mostly of bulk phases and only a little quantity carbonized matter. At very low heating rate of 2°C/min, the MOF precursor was completely carbonized, and no bulk phase was observed in the SEM images as shown in Figure 3.4. This result was indeed supported by the results of BET analysis. The sample with low heating rate showed a high BET surface area of 889 m^2/g. Hence it can be concluded that low heating rates will improve the porosity and prevent the agglomeration of particles. Moreover, low heating rates will increase the carbonization time and lead to evaporation of residual organic matter and unstable species.

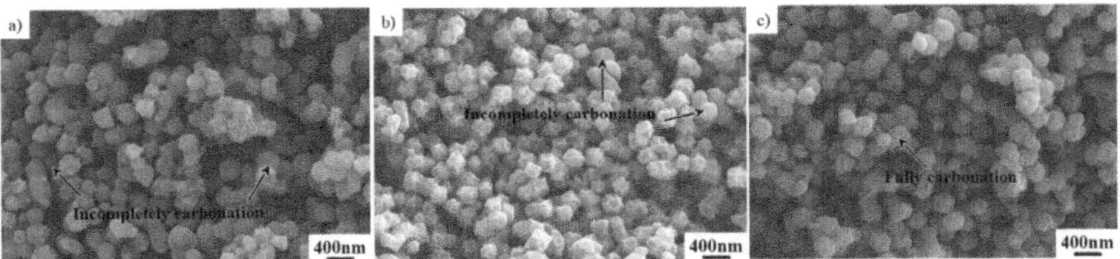

FIGURE 3.4   SEM images of Co-ZIF-8 pyrolyzed at different heating rates (a) 10°C/min, (b) 5°C/min, and (c) 2°C/min. Reproduced with permission from Ref. [32].

3.4.4   EFFECT OF CARBONIZATION TIME

Heating times will also affect the characteristic features of MOF-derived materials such as surface area, pore size, particle size and the composition. In one of the reports where pyrolysis was done in the presence of NH_3 source, the increase in pyrolysis time improved the defective sites in the resulting matrix due to the increased etching process [33]. Pyrolysis time also influences the pore size of many MOF-derived materials and also the particle size of the nano particles. Other studies showed that the increase in pyrolysis time favoured the orderly and uniform growth of particles since enough time was provided for the diffusion of the species.

3.5   STRATEGIES INVOLVED IN ENGINEERING THE MOF STRUCTURE

3.5.1   SILICA AS A SOURCE

Silica (SiO_2), being a cheapest non-toxic material, is widely used in the synthesis of MOF-derived nanomaterials as a template or structure directing agent, protective shell for preventing the catalyst from degradation. Silica prevents the particles from aggregating during thermal transformation and results in the formation of hierarchical porous structure. Employing SiO_2 as a template helped in thwarting the inward volume contraction of ZIF-8 particles during pyrolysis process [34]. The carbonization process was observed to start at the interface of ZIF-8 and SiO_2. Using an appropriate thickness of SiO_2 resulted in the formation of hollow structures due to the outward contraction of ZIF-8. The SiO_2 shell pulled the ZIF-8 particles towards itself, and the carbonization happened at the rigid interface. A thin SiO_2 shell resulted in the formation of material without any cavities. Silica not only affected the volume contraction but also impacted thermal stability. Other instances were also reported where the addition of SiO_2 helped in the prevention of agglomeration of particles. Thus, SiO_2 was found to be one of the viable sources to regulate the characteristics of the derived materials.

3.5.2   POLYMER AS A SOURCE

Polymers can be a potential candidate for employing it as a template acting not only as structure-directing agent but also as a source of carbon. Polymers can easily provide home to many guest species or doped contents. Due to the presence of π-π interactions and van der Walls forces, they can be efficiently coated on the surface of nano particles. Polymers such as resorcinol-formaldehyde (RF) [35], polydopamine (PDA) [36] had been extensively applied in the pyrolysis of ZIF-8 to obtain hollow carbon structures. Similar to the mechanism explained in above section, here also, the stress-induced orientation volume shrinkage plays an important role in obtaining hollow structures. Core-shell structures were also obtained using polymers as a template and ZIF-67 as the MOF precursor. A mixture of $Co(NO_3)_2$ and 4-aminophenol were added to the solution of formaldehyde and the organic linker 2-methyl imidazole. Due to the differences in the kinetics of formation of ZIF-67 and that of polymerization, a core-shell shell architecture is formed. After pyrolysis led to the formation of N/Co-doped porous carbon core-shell structure [37].

3.5.3 Surfactant Approach

There are wide variety of anionic and cationic surfactants such as cetyltrimethylammonium bromide (CTAB), sodium dodecyl sulphate (SDS), and Pluronic F-127 which can regulate the morphology of the MOFs during the pyrolysis process. These surfactants help in the prevention of agglomeration of nanoparticles (NP) during pyrolysis. They tend to form micelles protecting the NPs from aggregation. Moreover, size of the NPs can also be controlled with the addition of surfactants. Likewise, many other surfactants were explored to synthesize various single metal atom doped porous carbon structures.

3.5.4 Sonochemistry

Ultrasonic-assisted synthesis of derived structures was another area explored where morphological control was successfully attained. In one of the works reported by Xu et al., post carbonization KOH treatment was done using ultrasonic energy [38]. During the KOH treatment, ultrasonic energy assisted in the movement of K^+ ions into the spacing of graphene layers. Further thermal treatment helped in the formation of K_2CO_3 and subsequent decomposition of K_2CO_3 resulted in the formation of graphene nano ribbons with multiple layers. The reaction of K_2CO_3 with carbon occurred during the heat treatment to give 2–6 layered graphene nanoribbons. The resulting material exhibited good quality which was due to the KOH etching assisted by ultrasonication.

3.5.5 Salt Template Approach

In this method, salts are added as templates to assist the formation of various morphologies. Salts like NaCl are added especially to ZIF structures which show strong interactions due to their low energy C-N bonds. 2D nanosheets, web-like structures can be prepared with the addition of NaCl as a stable structure directing agent. Zhao and his group reported a web-like porous structure with NaCl assisted method using ZIF-67 as MOF precursor [39]. During the thermal transformation process, a web-like structure was formed upon the molten salt incorporation and the morphology was preserved even after pyrolysis. Similar experiment was conducted using other MOFs as precursors such as MOF-74, HKUST and UiO-66 by the same group. But none of them retained the web-like morphology after pyrolysis due to the presence of strong C-O bonds. Likewise, an intense research is going to fabricate new kinds of porous structures employing various salts.

3.6 CONCLUSION

MOF, as emerging category of nanomaterials, can potentially serve as a template than the traditional templates for obtaining porous carbon-based nanomaterials and other kinds of nanocomposite structures. Due to their interconnected structure and pore optimization and feasibility of external doping, several characteristic features can be imparted to the derived materials upon thermal treatment. Besides, the morphology transformation is an additional advantage with MOF materials. Optimizing the morphology can alleviate specific properties for targeted applications. The conductivity, charge transfer resistance, mass and charge transfer and other opto-electronic properties can be greatly improved by modifying the morphology of the derived materials which is not possible using traditional methods. Morphology can greatly affect the specific surface area, pore size distribution, heteroatom distribution and prevents the agglomeration of NP during the pyrolysis process. Hence, special attention must be given to control the morphology during the thermal treatment to impart specific features to the resulting materials for applications such as adsorption, catalysis, sensing, separation, energy storage and other emerging areas.

Morphology can be affected by several factors starting from the selection of MOF precursor, pyrolysis conditions and post-treatment methods. MOF precursors such as HKUST, ZIF, UiO, MOF-5, MOF-74 exhibiting different morphologies can be used as starting materials. Each kind of MOF is suitable for specific application and careful selection of the precursor is of utmost important. Pyrolysis conditions such as temperature, time, gaseous environment and external sources plays a crucial role in controlling the morphology. Altering these condition may result in specific structures and sometimes exceeding the limit of these conditions may result in aggregation of NP or collapse of the structures. Precise control of these features will contribute to the morphological transformation.

Templating strategies, either self-templating (using MOF alone as precursor) or external templating (addition of external agents such as SiO_2, polymers and other sources) can assist in the formation of specific morphologies. They can help in directional structural formation or can assist in the protection of NP from agglomeration. Hence, these sources must be utilized in order to obtain targeted results. To conclude, morphology is one of the important parameters that play a vital role in the formation of MOF-derived nanomaterials. Careful selection of the parameters and employing required external agents can give specific morphology to the material and can enhance the properties of the material.

REFERENCES

1. Wang, C.; Kim, J.; Tang, J.; Kim, M.; Lim, H.; Malgras, V.; You, J.; Xu, Q.; Li, J.; Yamauchi, Y. New Strategies for Novel MOF-Derived Carbon Materials Based on Nanoarchitectures. *Chem.* Elsevier Inc January 9, 2020, 19–40. https://doi.org/10.1016/j.chempr.2019.09.005.

2. Dang, S.; Zhu, Q. L.; Xu, Q. Nanomaterials Derived from Metal-Organic Frameworks. *Nat Rev Mater.* Nature Publishing Group December 5, 2017. https://doi.org/10.1038/natrevmats.2017.75.

3. You, B.; Jiang, N.; Sheng, M.; Gul, S.; Yano, J.; Sun, Y. High-Performance Overall Water Splitting Electrocatalysts Derived from Cobalt-Based Metal–Organic Frameworks. *Chem Mater*, 2015, *27* (22), 7636–7642. https://doi.org/10.1021/acs.chemmater.5b02877.

4. Lin, Q.; Bu, X.; Kong, A.; Mao, C.; Zhao, X.; Bu, F.; Feng, P. New Heterometallic Zirconium Metalloporphyrin Frameworks and Their Heteroatom-Activated High-Surface-Area Carbon Derivatives. *J Am Chem Soc*, 2015, *137* (6), 2235–2238. https://doi.org/10.1021/jacs.5b00076.

5. Wu, H. Bin; Xia, B. Y.; Yu, L.; Yu, X. Y.; Lou, X. W. Porous Molybdenum Carbide Nano-Octahedrons Synthesized via Confined Carburization in Metal-Organic Frameworks for Efficient Hydrogen Production. *Nat Commun*, 2015, *6*. https://doi.org/10.1038/ncomms7512.

6. Chen, Y. Z.; Wang, C.; Wu, Z. Y.; Xiong, Y.; Xu, Q.; Yu, S. H.; Jiang, H. L. From Bimetallic Metal-Organic Framework to Porous Carbon: High Surface Area and Multicomponent Active Dopants for Excellent Electrocatalysis. *Adv Mater*, 2015, *27* (34), 5010–5016. https://doi.org/10.1002/adma.201502315.

7. Xu, H.; Zhou, S.; Xiao, L.; Wang, H.; Li, S.; Yuan, Q. Fabrication of a Nitrogen-Doped Graphene Quantum Dot from MOF-Derived Porous Carbon and Its Application for Highly Selective Fluorescence Detection of Fe3+. *J Mater Chem C Mater*, 2015, *3* (2), 291–297. https://doi.org/10.1039/c4tc01991a.

8. Fang, L.; Xie, Y.; Wang, Y.; Zhang, Z.; Liu, P.; Cheng, N.; Liu, J.; Tu, Y.; Zhao, H.; Zhang, J. Facile Synthesis of Hierarchical Porous Carbon Nanorods for Supercapacitors Application. *Appl Surf Sci*, 2019, *464*, 479–487. https://doi.org/10.1016/j.apsusc.2018.09.124.

9. Jiang, W.; Pan, J.; Liu, X. A Novel Rod-Like Porous Carbon with Ordered Hierarchical Pore Structure Prepared from Al-Based Metal-Organic Framework without Template as Greatly Enhanced Performance for Supercapacitor. *J Power Sources*, 2019, *409*, 13–23. https://doi.org/10.1016/j.jpowsour.2018.10.086.

10. Oh, H. G.; Park, S.-K. Co-MOF Derived MoSe2@CoSe2/N-Doped Carbon Nanorods as High-Performance Anode Materials for Potassium Ion Batteries. *Int J Energy Res*, 2022, *46* (8), 10677–10688. https://doi.org/https://doi.org/10.1002/er.7866.

11. Li, Y.; Li, W.; Yang, C.; Tao, K.; Ma, Q.; Han, L. Engineering Coordination Polymer-Derived One-Dimensional Porous S-Doped Co3O4 Nanorods with Rich Oxygen Vacancies as High-Performance Electrode Materials for Hybrid Supercapacitors. *Dalton Trans*, 2020, *49* (30), 10421–10430. https://doi.org/10.1039/D0DT02029J.

12. Su, P.; Xiao, H.; Zhao, J.; Yao, Y.; Shao, Z.; Li, C.; Yang, Q. Nitrogen-Doped Carbon Nanotubes Derived from Zn–Fe-ZIF Nanospheres and Their Application as Efficient Oxygen Reduction Electrocatalysts with in Situ Generated Iron Species. *Chem Sci*, 2013, *4* (7), 2941–2946. https://doi.org/10.1039/C3SC51052B.

13. Lianli, Z.; Yong-Sheng, W.; Chun-Chao, H.; Miao, W.; Yu, W.; Hao-Fan, W.; Zheng, L.; Qiang, X. One-Step Synthesis of Ultrathin Carbon Nanoribbons from Metal–Organic

Framework Nanorods for Oxygen Reduction and Zinc–Air Batteries. *CCS Chem*, 2021, *4* (1), 194–204. https://doi.org/10.31635/ccschem.021.202101160.

14. Bhosale, R.; Bhosale, S.; Kumbhar, P.; Narale, D.; Ghaware, R.; Jambhale, C.; Kolekar, S. Design and Development of a Porous Nanorod-Based Nickel-Metal–Organic Framework (Ni-MOF) for High-Performance Supercapacitor Application. *New J Chem*, 2023, *47* (14), 6749–6758. https://doi.org/10.1039/D3NJ00456B.

15. Pang, H.; Guan, B.; Sun, W.; Wang, Y. Metal-Organic-Frameworks Derivation of Mesoporous NiO Nanorod for High-Performance Lithium Ion Batteries. *Electrochim Acta*, 2016, *213*, 351–357. https://doi.org/10.1016/j.electacta.2016.06.163.

16. Zhang, X.; Ou-Yang, W.; Zhu, G.; Lu, T.; Pan, L. Shuttle-Like Carbon-Coated FeP Derived from Metal-Organic Frameworks for Lithium-Ion Batteries with Superior Rate Capability and Long-Life Cycling Performance. *Carbon N Y*, 2019, *143*, 116–124. https://doi.org/10.1016/j.carbon.2018.11.005.

17. Cai, Y.; Fang, G.; Zhou, J.; Liu, S.; Luo, Z.; Pan, A.; Cao, G.; Liang, S. Metal-Organic Framework-Derived Porous Shuttle-Like Vanadium Oxides for Sodium-Ion Battery Application. *Nano Res*, 2018, *11* (1), 449–463. https://doi.org/10.1007/s12274-017-1653-9.

18. Xu, X.; Li, J.; Wang, M.; Liu, Y.; Lu, T.; Pan, L. Shuttle-Like Porous Carbon Rods from Carbonized Metal–Organic Frameworks for High-Performance Capacitive Deionization. *ChemElectroChem*, 2016, *3* (6), 993–998. https://doi.org/10.1002/celc.201600051.

19. Chang, G.; Zhang, H.; Yu, X. Y. 2D Metal–Organic Frameworks and Their Derivatives for the Oxygen Evolution Reaction. *J Alloys Compd*. Elsevier Ltd October 25, 2022. https://doi.org/10.1016/j.jallcom.2022.165823.

20. Lu, C.; Li, Z.; Xia, Z.; Ci, H.; Cai, J.; Song, Y.; Yu, L.; Yin, W.; Dou, S.; Sun, J.; et al. Confining MOF-Derived SnSe Nanoplatelets in Nitrogen-Doped Graphene Cages via Direct CVD for Durable Sodium Ion Storage. *Nano Res*, 2019, *12* (12), 3051–3058. https://doi.org/10.1007/s12274-019-2551-0.

21. Lin, Y.; Wan, H.; Chen, F.; Liu, X.; Ma, R.; Sasaki, T. Two-Dimensional Porous Cuprous Oxide Nanoplatelets Derived from Metal-Organic Frameworks (MOFs) for Efficient Photocatalytic Dye Degradation under Visible Light. *Dalton Trans*, 2018, *47* (23), 7694–7700. https://doi.org/10.1039/c8dt01117f.

22. Yang, S. J.; Im, J. H.; Kim, T.; Lee, K.; Park, C. R. MOF-Derived ZnO and ZnO@C Composites with High Photocatalytic Activity and Adsorption Capacity. *J Hazard Mater*, 2011, *186* (1), 376–382. https://doi.org/10.1016/j.jhazmat.2010.11.019.

23. Su, Y.; Li, S.; He, D.; Yu, D.; Liu, F.; Shao, N.; Zhang, Z. MOF-Derived Porous ZnO Nanocages/RGO/Carbon Sponge-Based Photocatalytic Microreactor for Efficient Degradation of Water Pollutants and Hydrogen Evolution. *ACS Sustainable Chem Eng*, 2018, *6* (9), 11989–11998. https://doi.org/10.1021/acssuschemeng.8b02287.

24. Cao, X.; Zheng, B.; Rui, X.; Shi, W.; Yan, Q.; Zhang, H. Metal Oxide-Coated Three-Dimensional Graphene Prepared by the Use of Metal–Organic Frameworks as Precursors. *Angew Chem Int Ed*, 2014, *53* (5), 1404–1409. https://doi.org/10.1002/anie.201308013.

25. Li, Z.; Shao, M.; Zhou, L.; Zhang, R.; Zhang, C.; Wei, M.; Evans, D. G.; Duan, X. Directed Growth of Metal-Organic Frameworks and Their Derived Carbon-Based Network for Efficient Electrocatalytic Oxygen Reduction. *Adv Mater*, 2016, *28* (12), 2337–2344. https://doi.org/10.1002/adma.201505086.

26. Oar-Arteta, L.; Wezendonk, T.; Sun, X.; Kapteijn, F.; Gascon, J. Metal Organic Frameworks as Precursors for the Manufacture of Advanced Catalytic Materials. *Mater Chem Front*, 2017, *1* (9), 1709–1745. https://doi.org/10.1039/C7QM00007C.

27. Zhao, P.; Hua, X.; Xu, W.; Luo, W.; Chen, S.; Cheng, G. Metal–Organic Framework-Derived Hybrid of Fe3C Nanorod-Encapsulated, N-Doped CNTs on Porous Carbon Sheets for Highly Efficient Oxygen Reduction and Water Oxidation. *Catal Sci Technol*, 2016, *6* (16), 6365–6371. https://doi.org/10.1039/C6CY01031H.

28. Li, Q.; Xu, P.; Gao, W.; Ma, S.; Zhang, G.; Cao, R.; Cho, J.; Wang, H.-L.; Wu, G. Graphene/Graphene-Tube Nanocomposites Templated from Cage-Containing Metal-Organic Frameworks for Oxygen Reduction in Li–O2 Batteries. *Adv Mater*, 2014, *26* (9), 1378–1386. https://doi.org/10.1002/adma.201304218.

29. Deng, A.; Yin, Y.; Liu, Y.; Xu, Y.; He, H.; Yang, S.; Qin, Q.; Sun, D.; Li, S. Unlocking the Potential of MOF-Derived Carbon-Based Nanomaterials for Water Purification through Advanced Oxidation Processes: A Comprehensive Review on the Impact of Process Parameter Modulation. *Sep Purif Technol*. Elsevier B.V. August 1, 2023. https://doi.org/10.1016/j.seppur.2023.123998.

30. Li, H.; Liang, M.; Sun, W.; Wang, Y. Bimetal–Organic Framework: One-Step Homogenous Formation and Its Derived Mesoporous Ternary Metal Oxide Nanorod for High-Capacity, High-Rate, and Long-Cycle-Life Lithium Storage. *Adv Funct Mater*, 2016, *26* (7), 1098–1103. https://doi.org/10.1002/adfm.201504312.

31. Yang, P.; Song, X.; Jia, C.; Chen, H.-S. Metal-Organic Framework-Derived Hierarchical ZnO/NiO Composites: Morphology, Microstructure and Electrochemical Performance. *J Ind Eng Chem*, 2018, *62*, 250–257. https://doi.org/10.1016/j.jiec.2018.01.002.

32. Cui, N.; Bi, K.; Sun, W.; Wu, Q.; Li, Y.; Xu, T.; Lv, B.; Zhang, S. Effect of Pyrolysis Conditions on the Performance of Co–Doped MOF-Derived Carbon Catalysts for Oxygen Reduction Reaction. *Catalysts*, 2021, *11* (10). https://doi.org/10.3390/catal11101163.

33. Li, X.; Sun, Q.; Liu, J.; Xiao, B.; Li, R.; Sun, X. Tunable Porous Structure of Metal Organic Framework Derived Carbon and the Application in Lithium–Sulfur Batteries. *J Power Sources*, 2016, *302*, 174–179. https://doi.org/10.1016/j.jpowsour.2015.10.049.

34. Liu, C.; Huang, X.; Wang, J.; Song, H.; Yang, Y.; Liu, Y.; Li, J.; Wang, L.; Yu, C. Hollow Mesoporous Carbon Nanocubes: Rigid-Interface-Induced Outward Contraction of Metal-Organic Frameworks. *Adv Funct Mater*, 2018, *28* (6), 1705253. https://doi.org/https://doi.org/10.1002/adfm.201705253.

35. Yang, H.; Bradley, J. S.; Chan, A.; Waterhouse, I. N. G.; Nann, T.; Kruger, E. P.; Telfer, G. S. Catalytically Active Bimetallic Nanoparticles Supported on Porous Carbon Capsules Derived from Metal–Organic Framework Composites. *J Am Chem Soc*, 2016, *138* (36), 11872–11881. https://doi.org/10.1021/jacs.6b06736.

36. Wang, M. J.; Mao, Z. X.; Liu, L.; Peng, L.; Yang, N.; Deng, J.; Ding, W.; Li, J.; Wei, Z. Preparation of Hollow Nitrogen Doped Carbon via Stresses Induced Orientation Contraction. *Small*, 2018, *14* (52), 1804183. https://doi.org/10.1002/smll.201804183.

37. Zhang, M.; Wang, C.; Luo, R.; Zhang, W.; Chen, S.; Yan, X.; Qi, J.; Sun, X.; Wang, L.; Li, J. A Phenolic Resin-Assisted Strategy for MOF-Derived Hierarchical Co/N-Doped Carbon Rhombic Dodecahedra for Electrocatalysis. *J Mater Chem A Mater*, 2019, *7* (10), 5173–5178. https://doi.org/10.1039/C8TA10918D.

38. Pachfule, P.; Shinde, D.; Majumder, M.; Xu, Q. Fabrication of Carbon Nanorods and Graphene Nanoribbons from a Metal–Organic Framework. *Nat Chem*, 2016, *8* (7), 718–724. https://doi.org/10.1038/nchem.2515.

39. Qian, Y.; An, T.; Birgersson, K. E.; Liu, Z.; Zhao, D. Web-Like Interconnected Carbon Networks from NaCl-Assisted Pyrolysis of ZIF-8 for Highly Efficient Oxygen Reduction Catalysis. *Small*, 2018, *14* (16), 1704169. https://doi.org/10.1002/smll.201704169.

Part 2

Applications of MOF-Derived Materials

4 Adsorption and Storage

4.1 INTRODUCTION

Adsorption is a process that is confronted in our daily lives with the historical roots tracing back to the ancient period, yet it often goes overlooked. The ancient practices of adsorption witnessed the use of sand, wood charcoal, and clay for the purposes of desalination of water, clarification of fat and oil, and treatment of diseases [1]. Earlier, the sorption process was believed to be solely absorption. Later it was Heinrich Kayser who coined the term adsorption on the proposal by the physiologist Du Bois Reymond. After the first systematic quantitative adsorption study on the adsorption of air by charcoal by Scheele in 1773, the subsequent studies focused on various aspects of adsorption, including the decolorizing properties of charcoal, the exothermal nature of gas adsorption, quantification of the amount of gas adsorption in porous carbon, and wetting of solids by gases [2]. Adsorption is the process of purification or separation of one or more components from a fluid mixture onto the surface of a solid. The separation process by adsorption differs from absorption, in which the transfer of sorbate is limited to the surface of the adsorbent. It is, by definition, a process of concentrating one or more components from the bulk fluid mixture to the interface region between the solid adsorbent and bulk fluid. In contrast, in the absorption process, the sorbate is transferred to the bulk of the absorbent (Figure 4.1). The adsorption and absorption processes are proven to be the most promising technologies for removing pollutants, for instance, one of the major pollutants causing global warming, CO_2. However, the removal of the pollutants by absorption has limitations as the solvents used in the absorption process can be corrosive, noxious, inflammable, or toxic. In addition, it involves high energy requirements for solvent regeneration. Adsorption can overcome the innate limitations of absorption. Table 4.1 compares the performance criteria of these processes. In the case of adsorption, trace impurities can be removed from the gas or liquid mixtures aiming at purification or separation. The process of trace impurity removal without the recovery of the adsorbed component is known as purification, while the impurity removal with the recovery of the adsorbate resulting in two or more streams, enriched with the valuable component, is known as bulk separation. Adsorption may demand high capital costs for industrial applications but has low operating costs due to its energy efficiency.

DOI: 10.1201/9781003432357-6

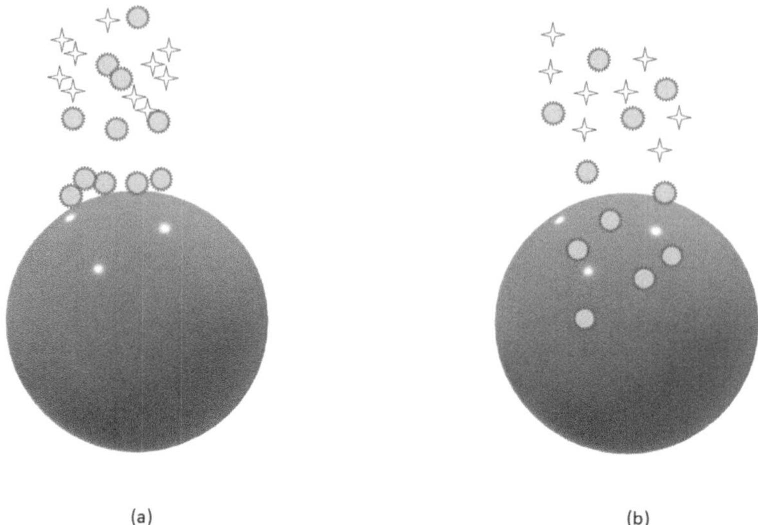

FIGURE 4.1 Selective removal of a sorbate by (a) adsorption and (b) absorption.

TABLE 4.1
Comparison of adsorption and absorption processes

Parameter	Adsorption	Absorption
Phenomenon	Surface phenomenon	Bulk phenomenon
Physical principle	Separation is by means of selective adhesion of an adsorbate molecule on the surface of the adsorbent	Selective dissolution of a solute molecule into the bulk of the solvent
Factors influencing the choice of adsorbent/solvent	Surface area of adsorbent, porosity, selectivity, functionality, capacity, life.	Solubility of the solvent, Solvent should be low volatile, non-toxic, non-corrosive, non-noxious, non-inflammable, cheap, chemically stable for easy regeneration,
Equilibrium data	Adsorption isotherms (Type I to Type VI)	Solubility data, vapour pressure data, equilibrium distribution coefficient data
Regeneration	Thermal swing, pressure swing, vacuum swing adsorptions, and displacement desorption	High temperature for chemical solvent, depressurization for physical solvent, adsorption, distillation

The adsorption process can even happen at a pressure less than the vapour pressure. The accompanied change in enthalpy of the process is given by

$$\Delta H = \Delta G + T\Delta S \tag{4.1}$$

where H is enthalpy, G is Gibbs free energy, T is temperature, and S is entropy. The degree of freedom in the adsorbed state decreases due to the association between the adsorbed molecule and the adsorbent surface. The translational degree of freedom of the molecules in the bulk gas phase is three, which is reduced to two when is adsorbed. Similarly, the rotational degrees of freedom also decrease for the molecule in the adsorbed phase. Therefore, a decrease in entropy with reduced randomness makes the change in entropy negative. As the adsorption process is spontaneous, the change in Gibbs free energy also becomes negative. Hence, the change in enthalpy is always negative, and the process is always exothermic. Initially, the magnitude of ΔH is greater than $T\Delta S$. However, as the adsorption progresses, ΔH becomes less negative and ultimately it becomes equal to $T\Delta S$ leading to $\Delta G = 0$, which corresponds to the equilibrium condition.

4.2　TYPES OF ADSORPTION

The surface force field of the adsorbent can be either physical or chemical, causing the selective adhesion of a component on the surface due to the different affinity towards the various components. The two types of adsorption, namely physical adsorption and chemisorption, are classified based on the energetics between the adsorbent and adsorbate. Physical adsorption occurs due to the weak forces between adsorbent and adsorbate while the strong chemical bond between them causes chemisorption.

4.2.1　PHYSICAL ADSORPTION

Physisorption is a reversible phenomenon in which the adsorbate gas molecules exhibit more vital intermolecular forces of attraction with the solid adsorbent surface than among the gaseous molecules, which in turn causes condensation of the gas molecules. The adsorbents are economical only when they are readily available for reuse through the desorbing mechanism of the adsorbed gas molecules from their surface. Regeneration of adsorbent is achieved by either decreasing the partial pressure of the bulk gas phase or increasing the temperature. The energy required to regenerate the adsorbent with physisorbed molecules is usually less than that of the chemisorbed molecule as the heat of adsorption is less (typically in the range of 20–30 kJ/mol). Moreover, the rate of physisorption is higher than that of chemisorption owing to the lower activation energy requirement. Usually, adsorption does not involve the dissolution of a gas molecule into the solid surface or penetration into the crystal lattice but is limited to the surface of the solid adsorbent. However, the gas molecules can diffuse through the pores and get adsorbed on the surface of the pores.

The adsorption process can be explained with the help of the potential energy surface shown in Figure 4.2. The dips on the surface with varying depths correspond to the adsorption sites, the depths of which represent the different energy levels. When

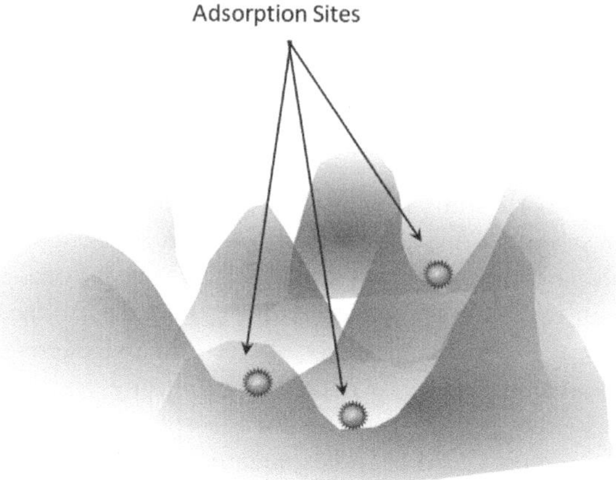

FIGURE 4.2　Potential energy surface.

a gas molecule collides elastically with the sorbent surface, it goes back to the gas phase without getting adsorbed on the sorbent surface as there is no energy gain or loss. On the contrary, during the inelastic collisions of gas molecules with the sorbent surface, they lose their kinetic energy leading to adsorption. Physisorption usually involves weak interactions, like van der Walls forces, that happens at the shallow dips on the potential energy surface as shown in Figure 4.2. In this case, the gas molecules undergo surface diffusion across the wells and get adsorbed on the adsorption sites. They remain adsorbed until they gain energy to escape into the gas phase. In the form of intermolecular potential, the weak interactive, van der Waals force between the molecules drops with the distance between molecules. The attractive potential between the molecules is signified by the second virial coefficient B(T) of the equation of state given below.

$$\frac{PV}{nRT} = 1 + \frac{B(T)}{V} + \frac{C(T)}{V^2} + \frac{D(T)}{V^3} + \dots \tag{4.2}$$

The Lennard-Jones potential, describing the potential energy of two interacting atoms or molecules as a function of the distance between them as given in equation 4.3 is demonstrated in Figure 4.3.

$$\varnothing^{LJ} = 4\varepsilon\left(\left(\frac{\sigma}{r}\right)^{12} - \left(\frac{\sigma}{r}\right)^{6}\right) \tag{4.3}$$

where, $\varnothing^{LJ}$ is intermolecular potential between atoms or molecules.

ε is bond energy indicating the potential well depth.

σ is bond length which is also referred to as the van der Waals radius.

r is the distance of separation between the two atoms or molecules

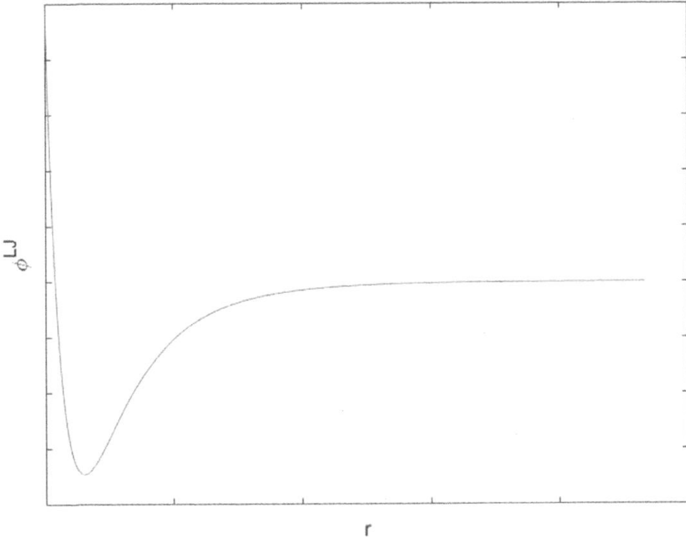

FIGURE 4.3 Lennard-Jones Potential plot.

The potential energy is the result of the two competitive forces namely (i) the long-range attractive force of order r^{-6} between the two interacting molecules and (ii) the short-range repulsive force of order r^{-12}, defined by Pauli's exclusion principle. The long-range attractive forces include dipole-dipole, dipole-induced dipole, and London forces of r^{-6}, r^{-8}, and r^{-12}, respectively. Truncating the higher-order dispersion terms, we obtain the well-known Lennard Jones potential equation.

For the interaction between two different molecules, the geometric and arithmetic mean of Lennard Jones parameter, ε and σ, respectively, can be used. The difference in the polarity between the sorbent and adsorbate or the surface reactivity is the reason for the selective interaction of the sorbent surface with the adsorbate. The porosity of the sorbent surface plays a major role in determining the adsorption capacity of sorbent material.

4.2.2 CHEMISORPTION

Chemisorption is caused by chemical interaction between the adsorbed molecule and the solid surface. The formation of the chemical bond in such interaction is often a result of the exchange, transfer, or sharing of electrons between them that leads to the formation of the chemical bond. This chemical interaction of the solid surface with the adsorbate molecule is highly selective. Unlike physisorption, chemisorption forms a monolayer due to the selective chemical interaction between the adsorbate molecule and the active sites on the adsorbent surface, as demonstrated in Figure 4.4, leading to the formation of a covalent bond. The strength of the chemical bond formed during chemisorption and the selectivity towards the adsorbate molecule depends on the composition, structure of adsorbents, and reactivity of

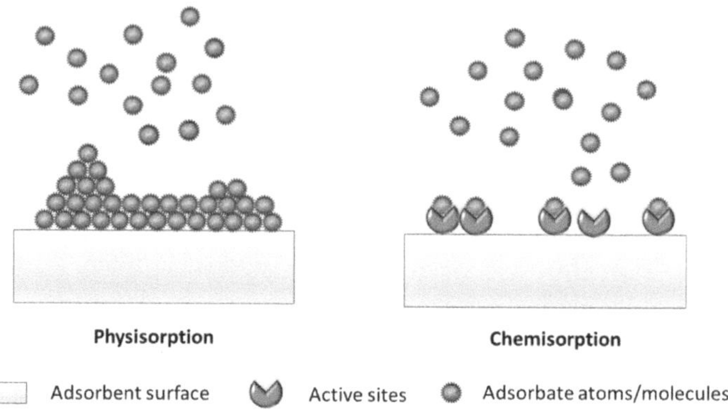

FIGURE 4.4   Schematic of physisorption and chemisorption.

TABLE 4.2
Physisorption vs chemisorption

	Physisorption	Chemisorption
Temperature	Occurs at a lower temperature than the chemisorption typically near or below the condensation point of gas. Adsorption decreases with increasing temperature	Effective absorption is limited to a narrow range of temperatures though it can occur over a wide range of temperatures
Adsorption enthalpy	Narrow range, ranging between 5 and 40 kJ/mol	Wider range, ranging between 40 and 800 kJ/mol
Nature of interaction	Weak interaction forces like van der Waals force, reversible, nonspecific	Strong chemical bond, irreversible, specific
Adsorption phenomenon	Multilayer adsorption	Monolayer adsorption
Activation energy	Low activation energy, nonactivated process	High activation energy, activated process

adsorbents with adsorbate molecules. The difference between physisorption and chemisorption is presented in Table 4.2. It should be noted that although no new chemical compounds are generated, they form covalent bonds of different strengths, much greater than that in physical adsorption. The bond strength only determines the energy required for regeneration of the adsorbent, which is of the order of heat of reaction that can cause the chemical modification of the adsorbent. The high regeneration energy requirement makes this process irreversible. Chemisorption is of primary importance in the catalytic reaction.

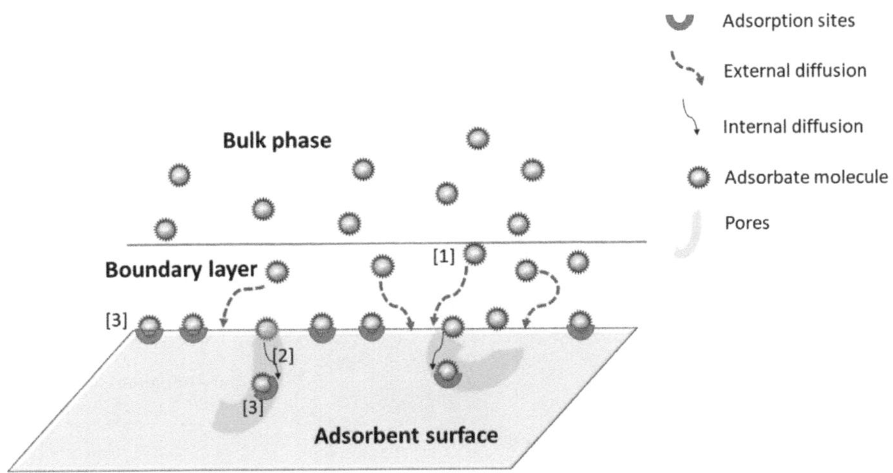

FIGURE 4.5   Steps involved in Adsorption [1] External diffusion [2] Internal diffusion [3] adsorption of molecules on adsorption sites available at the surface and the pore walls.

4.3   STEPS INVOLVED IN ADSORPTION

Adsorption of a molecule involves the following three consecutive steps, as shown in Figure 4.5. Firstly, the adsorbate molecules must reach the solid surface from the bulk fluid phase by an external diffusion process, the rate of which is proportional to the concentration difference of adsorbate between the bulk and the solid surface. Secondly, the adsorbate molecules access the adsorption sites available inside the pores through internal diffusion. Lastly, the adsorption of the molecules happens on the adsorption sites available at the surface and inside the pores.

4.4   TYPES OF PORES IN A POROUS ADSORBENT

The porous adsorbents exhibit different types of pores which can broadly be classified into 'closed pores' and 'open pores'. The schematic of a cross-section of a typical porous adsorbent is depicted in Figure 4.6. Closed pores are the ones that are completely isolated from the neighbouring pores and do not have any access to the external fluid. They do not facilitate the transport of fluid through the pores and are not very useful for the adsorption of gases as they do not have access to the external fluid. On the contrary, the open pores have channels with access to the external fluid and allow the transport of fluid through them. The open channels can further be classified as through channels and blind channels. The through channels have the pore opening starting at one location and ending at a different location while the latter has an opening at only one end with the other end closed. In addition, the porous adsorbent can exhibit an external surface with undulation. Depending on the width and depth of the undulation it can either be called roughness or blind pore. The wider surface irregularities are considered as surface roughness while the deeper irregularities are treated as blind pores.

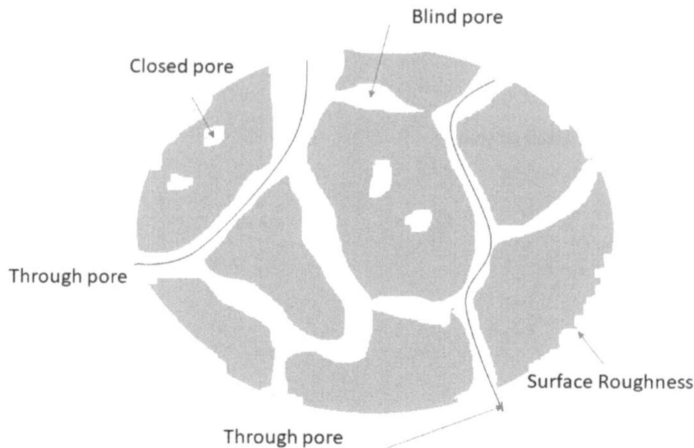

FIGURE 4.6 Schematic of a cross-section of a porous adsorbent.

4.5 ADSORPTION ISOTHERM

The maximum adsorption capacity of a porous material can never be realized completely due to the resistances offered during various steps of adsorption as illustrated in Figure 4.5. Therefore, to estimate the practical adsorption capacity of an adsorbent, the information on adsorption equilibrium is essential. The amount of adsorbate molecules in adsorbed form is usually expressed as mass adsorbed per unit mass of adsorbent, while in the fluid phase is expressed as mole percent or partial pressure. The amount adsorbed usually depends on the concentration of adsorbate molecules available in the fluid phase, which is a function of temperature. At any given temperature an equilibrium condition exists between the concentration of adsorbate on the solid surface and the fluid phase, which is represented as an adsorption isotherm. The amount of adsorption increases with decreasing temperature and increasing pressure as shown in Figure 4.7, each curve representing the adsorption isotherm at different temperatures. The shape of the isotherms is not only influenced by temperature and pressure but also by the nature of the adsorbent surface and adsorbate molecules. For example, gases like ammonia, sulphur dioxide, or carbon dioxide are more readily adsorbed due to the greater van der Walls forces than other gases like oxygen, nitrogen, or hydrogen. Similarly, the adsorbents with larger surface areas can adsorb more adsorbates. Furthermore, the pore size of the adsorbent promotes different adsorption behaviour and thus influences the shape of the adsorption isotherm. The adsorption behaviour of the porous adsorbents with different sizes are presented in Table 4.3. Pores with widths of <2 nm are known as micropores, with the subcategories of (i) narrow or ultra micro pores, with a pore width of <0.7 nm in which fluid-pore wall interaction is the adsorption characteristic and (ii) wide or supermicropores, with a pore size of 0.7–2 nm in which fluid-pore wall interaction is dominating near the wall while at the centre it is governed by fluid-fluid interaction. Furthermore, the pore sizes of 2–50 nm are classified as mesopores. Adsorption within the mesopores is characterized by fluid-fluid interaction that leads

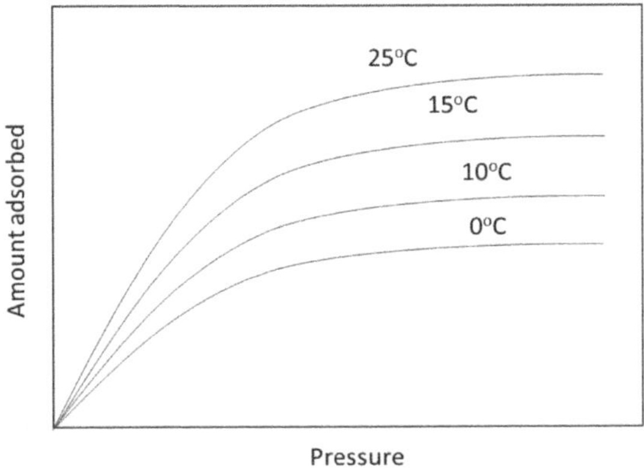

FIGURE 4.7   Adsorption capacity at different temperatures.

TABLE 4.3
Types of pores and their adsorption behaviour

Type of pore	Pore size	Adsorption characteristic
<2 nm	Micropore	Very narrow pore: Fluid–pore wall interaction dominates
		Slightly wider pore: Fluid–pore wall interaction near the wall, fluid-fluid interaction at the pore centre
2–50 nm	Mesopore	Fluid-fluid interaction resulting in multilayer adsorption and capillary condensation
		Factors affecting the stability of the multilayer film change with the thickness of the film.
		Thin film: Adsorption potential
		Thicker film: surface tension and curvature effect
>50 nm	Macropore	Fluid-pore wall interaction is insignificant, fluid density inside the pore is the same as bulk fluid density

to multilayer adsorption. The stability of the multilayer film is influenced by adsorption potential in the case of thin film while surface tension and curvature effect are the major influencing factors when the multilayer film goes thick. The pores with width >50 nm are categorized as macropores the adsorption of which is characterized by fluid-fluid interaction with the fluid density same as the bulk fluid density. Fluid-pore wall interaction becomes negligible in this case.

4.5.1   TYPES OF ADSORPTION ISOTHERMS

Gas adsorption isotherms are commonly classified into six major categories as summarized in Table 4.4. The first type of isotherm, Type I, is characterized by a reversible and concave nature, and it forms either a monolayer or a very few-layer adsorption. Type I(a) isotherm is limited to monolayer adsorption, which occurs due

TABLE 4.4
Types of adsorption isotherms

Type	Nature of adsorbent	Characteristics	Figure
Type-I(a)	Narrow microporous adsorbent (<1 nm width) The pore size in the range of molecular dimension	• Reversible isotherm • Accessible micropore volume limits the amount adsorbed • Steep increase in the amount adsorbed due to the micropore filling through the enhanced adsorbate-adsorbent interaction • Monolayer adsorption	I(a) A steep increase in the amount adsorbed
Type-I(b)	Adsorbent with wider micropore to narrow mesopore (1–2.5 nm)	• Reversible isotherm • Approaches limiting value with relatively less increase in the amount adsorbed at low relative pressure compared to Type-I(a)	I(b) Relatively less increase in the amount adsorbed compared to Type IA
Type II	Non-porous or macroporous adsorbent	• Reversible isotherm • Monolayer formation followed by multilayer adsorption up to high relative pressure • Sharp inflection point B – completion of monolayer coverage • Gradual inflection – the overlap of the completion of monolayer coverage and the beginning of multilayer adsorption • Thickness of multilayer adsorption keeps increasing and is not limited by the relative pressure	II B
Type-III	Non-porous or macroporous adsorbent	• Weak adsorbent-adsorbate interaction • Incomplete monolayer formation due to the assemblage of adsorbates on the selective adsorption sites • Limited adsorption at saturation pressure	III

TABLE 4.4 (Continued)
Types of adsorption isotherms

Type	Nature of adsorbent	Characteristics	Figure
Type-IV(a)	Mesoporous adsorbents with pore width greater than the critical width	• Both (i) the adsorbent–adsorbate interactions and (ii) the interaction among the adsorbate molecules are significant. • Formation of monolayer adsorption followed by multilayer adsorption. • Steep increase in amount adsorbed due to capillary condensation. • Multilayer adsorption is limited once the pores are filled. • Hysteresis due to larger pore width than the critical width which depends on the relative pressure.	
Type IV(b)	Mesoporous adsorbents with smaller pore width	• Similar to Type IVA isotherm but without hysteresis • Completely reversible isotherm • Similar to Type II isotherm but with limited multilayer adsorption	
Type V	Mesoporous adsorbent	• Relatively weak interaction between adsorbent and adsorbate and incomplete monolayer adsorption at a low relative pressure • Similar to Type III isotherm at a low relative pressure • At higher relative pressure condensation molecular clustering and pore filling	
Type VI	Highly uniform non-porous adsorbent	• Step-by-step adsorption • Layer by layer adsorption limited by the capacity of each adsorbed layer	

$$\theta_v = 1 - \theta_a \qquad (4.9)$$

On substitution of equation 4.9 in equation 4.7, we will get

$$\theta_a = \frac{Kp}{1 + Kp} \qquad (4.10)$$

where K is the adsorption equilibrium constant. According to the above equation, the Langmuir model explains the formation of a monolayer which is indicated by the $\theta_a \rightarrow$ 1 at high partial pressure and is proportional to partial pressure at low partial pressure.

The Langmuir model explains the Type-I adsorption isotherm where the isotherm is levelled off due to the monolayer adsorption. However, multilayer adsorption on the adsorbent surface is also common in which the initially adsorbed layer will act as the substrate for the subsequent layers as in the case of Type-II adsorption isotherm. Type-II isotherm keeps growing infinitely with increasing relative pressure as the newly formed adsorbed layer is acting as substrate for further adsorption. The multilayer adsorption can be explained by Brunauer-Emmett-Teller (BET) model given by equation 4.11.

$$\theta = \frac{cp}{\left(1 - p/p_0\right)(p_0 + p(c-1))} \qquad (4.11)$$

where θ is the ratio of the amount adsorbed to its monolayer equivalent, c is BET constant, p and p_0 are the pressure and vapour pressure.

4.6 MOF-DERIVED ADSORBENTS

MOFs are porous crystalline materials that attracted worldwide interest and showed rapid development in the past decade due to their variety of architecture, tunable porosity, and large specific surface area. Despite the favourable properties of MOFs for the adsorption and storage of carbon dioxide and hydrogen, it is crucial to address their limitations including poor adsorption capacity and selectivity towards CO_2 or H_2 uptake and poor thermal and aqueous stability. Recently, MOF materials have been explored for their potential use as templates for porous carbon materials characterized by enhanced adsorption capacities and better thermal and aqueous stabilities. Different types of MOF-derived materials including non-doped porous carbon, doped porous carbon, porous carbon decorated with metal/metal oxide, and nanoparticles are widely used in adsorption applications. The application of these different MOF-derived materials for adsorption/storage of CO_2 and H_2 will be discussed in the subsequent sections.

4.6.1 CO$_2$ ADSORPTION/STORAGE

The substantial increase in carbon dioxide emissions, a crucial greenhouse gas, stemming from anthropogenic activities has made a significant contribution to global warming due to its potential to trap heat in the atmosphere. The combustion of fossil fuels by the transport sector and industries in the modern era emits excessive CO_2 into the atmosphere thereby driving climate change. Controlling the emission of carbon through carbon capture and storage (CCS) is a more effective and easier way of stabilizing global warming compared to the control of emissions or reduction of other greenhouse gases. Though CCS has been commercially deployed, it still has the challenges of finding a suitable material for CO_2 capture and storage until it is transported to the storage site. The MOF-derived carbon material is one of the potential adsorbents that could offer a solution to this problem. The most commonly used adsorbents like activated carbon and zeolite have the limitations of weak interaction with CO_2. MOF-derived carbon can be more advantageous than these conventional adsorbents owing to their tunable properties through the control of carbonization parameters.

4.6.1.1 Porous Carbon for CO_2 Adsorption/Storage

A diverse range of mono- and bimetallic MOFs have been extensively investigated for carbon capture. Initially, the transition metal-based MOFs were subject to extensive research in the realm of carbon capture, thanks to their multivalent and high oxidation states. Subsequently, the search expanded to explore the other elements from the periodic table which demonstrated that except few MOFs like HKUST and MOF-5, the majority of the MOFs have poor thermal and aqueous stabilities. The ongoing research efforts are now dedicated to the development of adsorbent materials for the enhancement of adsorption capacity, and thermal and aqueous stabilities using MOF-based materials as a precursor to derive porous carbon with or without dopants and metal components and MOF-derived nanoparticles.

Porous carbon materials find wide applications in various fields, including gas storage and separation, catalysis, and energy owing to their large surface area, pore volume, and desirable pore size distribution, etc. Orderly porous carbon materials are commonly constructed by the templating method using mesoporous silica, aluminosilicates, or zeolites as the sacrificial templates along with some carbon precursors [4]. Compared to the template methods using conventional materials, the use of MOF as a template makes the process simple and efficient by eliminating the additional step of template removal. In addition, it does not require any additional carbon precursor as the organic linker of the MOF material itself can act as the carbon precursor. Indeed, the porosity of the parent MOF precursor will exert a substantial impact on the pore size of the MOF-derived carbon (MDC) material as well. The use of different organic ligands during the synthesis can result in MOFs with the same topography but with different pore sizes. The effect of change in the pore size of MOF material obtained from different organic ligands will be more predominant at high pressure compared to the adsorption at low pressure. Because, at high pressure, the pore size is sufficiently large, and hence more adsorbates can get adsorbed away from the linker molecules. For instance, hierarchical porous carbon (HPC) synthesized from different parent MOFs exhibits different textural properties and carbon capture performance [5]. The HPCs derived from MOF-5 and MOF-74 having the common metal part of Zn but different organic ligands including terephthalic acid, and 2,5-Dihydroxybenzene-1,4-dicarboxylic acid, respectively, exhibit different CO_2 adsorption capacity as given in Table 4.5. The difference in CO_2 uptake

TABLE 4.5

Illustrative MOF-derived materials for CO_2 adsorption

S.No	MOF-derived material	MOF precursor	Carbonization condition	Surface area (m²/g)	Pore volume (cm³/g)	Ultramicropore volume	Micropore volume (cm³/g)	CO_2 uptake (mmol/g)	Ref
1	C700W	ZIF-8	700°C, 3 h and subsequent washing with HCl	914	0.51		0.37	3.80 @298 K, 1 bar 5.51 @ 273 K, 1 bar	[7]
2	C800 NC800 BNC800-2	IL@MIL-100(Al) MIL-100(Al) composited with ionic liquids	800°C, 7h	2731 2397 1522	1.76 1.37 0.78			1.8 3.8 @298 K, 1 bar, 5.7 @273 K, 1 bar 3.2@298K, 1 bar 4.4@273K, 1 bar	[8]
3	HPC5b2-1000 HPC5b1-1000 HPC74-1000			2517 2462 1900	5.53 3.79 1.00			7.01 (PSA[a]) 3.93 (VSA[b]) 6.62 (PSA[a]) 3.79 (VSA[b]) 5.59 (PSA[a]) 3.88 (VSA[b])	[5]
4	nZDC700 nZDC1000	nZIF-8	700°C and 1000°C, 5°C/min, 6 h with subsequent HCl washing	950 1222			0.35 0.43	1.40 @ 0.15 bar 3.51@1 bar 7.82@20 bar 0.99@0.15 bar 3.22@1 bar 10.21@20 bar	[9]

5	MCU900	MOF-5	900°C, 4°C/min, 5 h	2307	2.54		0.26	3.71 ± 0.03 @0°C, 1bar 2.31 ± 0.02 @25°C, 1 bar	[10]
6	mJUC-160–900	mixed-ligand ZIF JUC-160	900°C, 5°C/min, 5 h	940	0.44	0.28	0.34	5.5 @273K, 3.5 @298 K	[11]
6	CZIF8a CZIF68a CZIF69a	ZIF-8(Zn(mIM)$_2$) ZIF-68(ZN(bIM) (nIM) ZIF-69(Zn(cbIM) (nIM))	1000°C, 5°C/min, 5 h followed by KOH activation	2437 1861 2264	1.266 1.086 1.184			4.04 0–1 atm, 273 K 4.49 0–1 atm, 273 K 4.76	[6]
7	NC-900	N-rich ZIF-8	900°C, 10°C/min, 8 h	2747	1.58			5.1 @273 K 3.9 @298K	[12]
8	Multiwall carbon nanotube (CN-1)	[Co(HTTG) (H$_2$O)$_2$]$_n$	600°C, 10°C/min, 1 h					1.4 @20°C, 1 bar	[13]
9	CF-500 CF-700	Zn-hfipbb	500°C and 700°C, 5oC/min, 3 h	561 774	0.33 0.46		0.19 0.23	1.8 @20°C 3.3@20°C	[14]
10	Porous carbon denoted as 'A'	ZIF-8	800°C, 3 h With 20 ml/min Ar atmosphere with saturated water vapour	1465	0.79		0.61	4	[7]

[a] PSA – Pressure swing adsorption (between 6 bar and 1 bar).
[b] VSA – Vacuum swing adsorption (between 1.5 bar and 0.5 bar).

at high pressure corresponding to pressure swing adsorption is more predominant compared to that corresponding to vacuum swing adsorption. In the same study, the influence of varying several factors including the precursor, solvent quantities and the treatment time on the textural properties and carbon capture capacity were also reported. The MOF-5 synthesized at different conditions, upon subsequent carbonization produced different HPCs namely HPC5b1 and HPC5b2 exhibiting different textural properties and CO_2 adsorption at pressure swing and vacuum swing adsorptions. Similar results have also been reported for the different zeolitic imidazole framework (ZIF) MOF materials synthesized employing the same Zn (II) metal ions but with different organic linker molecules. This led to the formation of MDCs with different local structures resulting in differences in the performance of CO_2 adsorption [6].

4.6.1.2 Doped MOF-Derived Porous Carbon for CO_2 Adsorption

The incorporation of nitrogen-containing functional groups into the MOF-derived porous carbon is beneficial for CO_2 adsorption as it introduces more basic adsorptive sites that can accommodate more acidic CO_2 molecules. MDC doped with nitrogen can be obtained by either carbonizing N-enriched MOF directly or with an additional external nitrogen source. During the carbonization of MOF or N-enriched MOF, though the crystallinity of the parent MOF material is lost, the morphology of the MDC material remains the same. Doping of MDC with nitrogen, boron, sulphur, or other metal or metal oxide nanoparticles depends on the type of the organic linker and the metal node present in the parent MOF material. Depending on the carbonization temperature, metal-free MDC or MDC hybrid with metal/metal oxide nanoparticles can be obtained. Generally, the surface area of metal-free MDC is higher than metal-doped MDC. The surface area of the ZIF-8 can be higher as it can result in nitrogen-doped metal-free MDC due to the low boiling point of zinc. MDC free from metal/metal oxide nanoparticles are the potential candidate for adsorptive removal of the contaminants while the MDC doped with the metal or metal oxides can be used for the catalytic applications. Metal-free MDCs can be used for the adsorptive removal of gas molecules from gas mixtures or the contaminants/hazardous chemicals from liquids.

Whether a secondary nitrogen source is employed in conjunction with the parent MOF or if MOF containing a nitrogen-rich organic ligand alone is used for nitrogen doping, the carbonization parameters and the choice of organic ligand used will influence the dopant level and the textural properties of the porous carbon derived from MOF. Direct carbonization of ZIF-8, ZIF-68, and ZIF-69 at a temperature of 1000°C for 5 h at a heating rate of 5°C/min followed by KOH activation resulted in MOF-derived porous carbons namely CZIF8a, CZIF68a, and CZIF69a, with surface area ranging from 1800 to 2400 cm^2/g [6]. The type-II nitrogen-sorption isotherm obtained indicates the presence of some large micropores as well. The BET surface area, pore volume, and nitrogen doping determine the carbon dioxide uptake capacity. The presence of residual nitrogen in the MDCs helps in enhancing the carbon dioxide uptake capacity. Carbon dioxide uptake is influenced the most by the local structure. The strong bipolar interaction between the polycyclic aromatic rings and carbon dioxide molecules makes CZIF68a and CZIF69a better candidates than CZIF8a.

More than surface area and the pore volume, the presence of more benzene-fused rings improves carbon capture. Nitrogen-doped porous carbon derived from ZIF8 at different carbonization temperatures from 600°C to 1000°C shows variation in its surface area, pore volume, nitrogen content, and subsequent carbon dioxide uptake capacity. Incorporation of nitrogen can be achieved by using a separate nitrogen source in addition to the nitrogen-containing organic ligand. Aijaz et al. synthesized N-rich ZIF-8 MOF using the secondary carbon precursor, furfuryl alcohol, and an additional source of nitrogen, NH_4OH, and calcined at different temperatures of 600°C, 700°C, 800°C, 900°C, and 1000°C for 8 h to obtain NC600, NC700, NC800, NC900, and NC1000, respectively [12]. Among the various nitrogen-enriched MOF-derived carbons studied, NC900 exhibited the most outstanding performance as an adsorbent for carbon capture. The surface area and pore volume increase with increasing carbonization temperature while the nitrogen content decreases. The highest surface area observed at a carbonization temperature of 1000°C does not correspond to the highest carbon capture, possibly owing to its lowest nitrogen content. Thus, it becomes apparent that CO_2 adsorption capacity is influenced not only by the textural properties but also by the dopant concentration within the porous carbon. Pure porous carbon derived from ZIF8 shows a monotonous increase in CO_2 uptake with increasing carbonization temperature. However, the subsequent washing and activation of the ZIF8-derived porous carbon exhibits a different trend with the highest CO_2 uptake corresponding to the sample carbonized at 700°C followed by HCl-washing. Both the textural properties including surface area, and pore volume, and the nitrogen content are the essential parameters determining the CO_2 uptake capacity. Surface polarity and basicity of the porous carbon increase due to the nitrogen content which actually favours the CO_2 adsorption. Most of the porous carbon derived from MOF material requires subsequent HCl washing for the removal of metal and metal oxides and KOH activation for better textural properties and adsorption capacity of MDCs. Surprisingly, subsequent KOH-activation of the HCl-washed porous carbon reduces the carbon uptake capacity due to the removal of nitrogen content. The decrease in the carbon uptake capacity during KOH activation can be addressed through K^+ ion exchange with the metal nodes of the MOF material.

Recently, it was reported that instead of a separate activation step subsequent to the carbonization of MOF, the introduction of K^+ ion exchange with the metal nodes of the parent MOFs facilitates the improved characteristics, thanks to the in-situ activation that occurs during the carbonization process itself [15]. Ion-exchanged MOF material was used as a potential precursor material to obtain activated porous carbon material. They synthesized K^+ exchanged N-doped carbon derived from Bio-MOF-1, an anionic framework with $MeNH^{2+}$ cations which were exchanged with K^+. The potassium ions introduced activate the porous carbon and introduce micropores compared to the parent Bio-MOF-1, which is largely mesoporous. Generally, the nitrogen content of the parent MOF is reduced during the carbonization process because of the high-temperature conditions maintained during carbonization. The incorporation of K^+ ions into the MOF structure helps in bringing down the carbonization temperature. During the carbonization of potassium ion-exchanged Bio-MOF-1, thermal decomposition of the MOF leads to the formation of potassium carbonate at lower temperatures and K_2O at higher temperatures. In addition, K will penetrate between the carbon layers formed during carbonization, and intercalate the layer, and increase the pore volume on its removal. Among the different carbonization conditions from 700°C to 1000°C, MDC carbonized at 700°C, with a large portion of micropore volume, and nitrogen content, demonstrated the highest carbon uptake capacity and selectivity towards nitrogen. The higher temperature causes over-activation by K^+ ions which causes an etching effect and reduces the surface area.

Other than nitrogen, the incorporation of heteroatoms like boron can significantly enhance the adsorption/storage capacity. Instead of using MOF with nitrogen-containing organic ligands as the precursor, an attempt was made by using a composite of nitrogen and boron-containing ionic liquids with MOF material [8]. MIL-100(Al) was combined with two ionic liquids namely 1-ethyl-3-methylimidazolium dicyanamide (EMIM-DCN) and 1-ethyl-3-methylimidazolium tetracyanoborate (EMIM-TCB) to incorporate nitrogen and boron heteroatoms into the composite, respectively. These composites were further carbonized at a temperature of 800°C for 7 h to obtain NC800 and BC800-1 (0.25 g of EMIM-TCB) and BC800-2 (0.125 g of EMIM-TCB). Pure MIL-100(Al) was also carbonized at 600°C, 800°C, and 1000°C to obtain C600, C800, and C1000, respectively. They were evaluated for their CO_2 uptake capacities. Compared to the C800, carbon derived from IL@MOF composite resulted in enhanced CO_2 uptake. The active basic centres available at IL@MOF-derived carbon acts as the ideal adsorption sites for CO_2 molecules due to their slightly acidic nature.

4.6.1.3 MOF-Derived Nanoparticles for CO_2 Adsorption

MOF material with a unique structure originating from the constituent metal element and organic ligand is an ideal sacrificial template for obtaining porous carbon also a precursor for obtaining the nanoparticles through the thermal transition process. Instead of MDC, the MOF-derived nanoparticles can also be employed as an adsorbent for carbon capture. Calcium looping is a process of capturing carbon in industries, especially in cement industries where CO_2 is removed from flue gases by using CaO as an adsorbent. Carbon capture by calcium looping involves two reversible reactions namely carbonization and calcination. Carbon captured during the carbonization of CaO resulting in the formation of $CaCO_3$ will release CO_2 on calcination. Despite the high theoretical adsorption capacity of CaO for CO_2 adsorption (17.86 mmol/g of CaO), it suffers from limitations, at higher temperatures, of reduced pore size and blocking of adsorption sites and aggregation of the CaO particles on sintering at high temperatures [16]. The incorporation of CaO particles into inert materials like Al, Si, or Zr can be helpful to overcome this issue to a certain extent [17, 18]. However, the interaction of Al, Si, and Zr with CaO will reduce the availability of the adsorption sites for CO_2 molecules. Fabrication of CaO in the nanoscale will minimize the diffusion length of CO_2 molecules and thus will enhance CO_2 uptake capacity. Very recently, MOF-derived CaO sorbent was reported for the high-temperature capture of industrial CO_2. Zixiao Liu et al. synthesized Ca-MOF using three different ligands namely lactic acid, 2-amino terephthalic acid, 1,3,5, benzene tricarboxylic acid, and used them as the precursor for nano CaO particles [19]. Among these three MOF precursors, CaO derived from Ca-MOF, lactic acid ligand resulted in the smallest nano

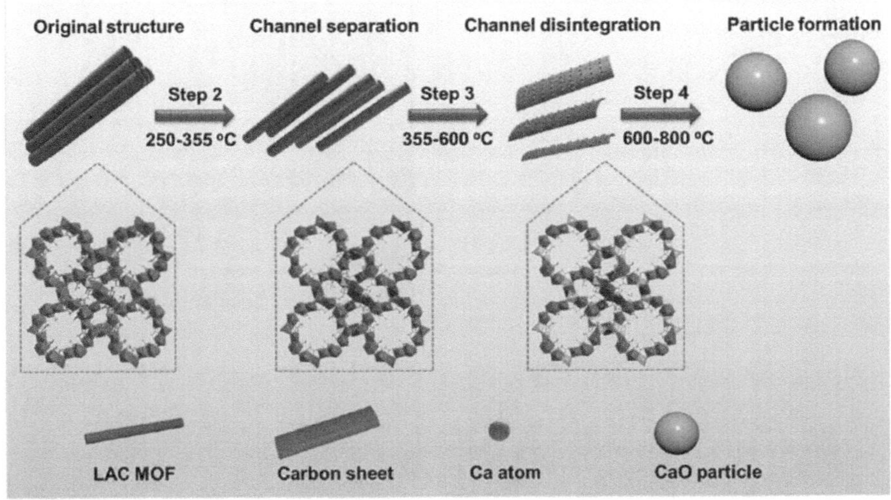

FIGURE 4.8 Structure transition from LAC-MOF to LAC-C. Reproduced with permission from Ref. [19].

CaO with an average size of 100 nm with a CO_2 adsorption capacity of 11.8 mmol/ g which demonstrated better stability even after four cycles. The adsorption capacity shows significant reduction with the other two ligands, 2-amino terephthalic acid and 1,3,5, benzene tricarboxylic acid mainly due to the particle sintering rather than the pore blocking. The thermal transition of lactic acid-Ca-MOF precursor to nano CaO through four major steps is depicted in Figure 4.8.

Thermogravimetric analysis and scanning electron microscopy analysis of the parent MOF and carbon sheet obtained during the intermediate step and the final CaO nanoparticles are depicted in Figure 4.9. After the removal of solvents and free ligands in the first step of initial temperature, the further increase in temperature removes the coordination units with increasing coordination number in the next three steps. Ca with low coordination number coordinated between the helical chains at the binding sites are unstable and are removed at relatively lower temperature while the Ca with high coordination number coordinated inside and outside the helical chain are removed at higher temperature. Removal of these low and high coordinated Ca causes the transformation of tube clusters to get separated and to open up the separated channels to form the carbon sheet structure, respectively. Further heating removes the carbon and forms the nano CaO particles. Nano CaO derived from Ca-MOF with the three different ligands shows no micropores in its structure. However, there is a reduction in CO_2 adsorption capacity with an increased number of cycles which is mainly due to the sintering effect rather than pore blocking.

4.6.1.4 MOF-Derived Multiwall Carbon Nanotube for CO_2 Adsorption

Removal of CO_2 is commercially achieved through alkali absorption or by adsorption through a porous medium. To combine the benefit of both alkali and porous

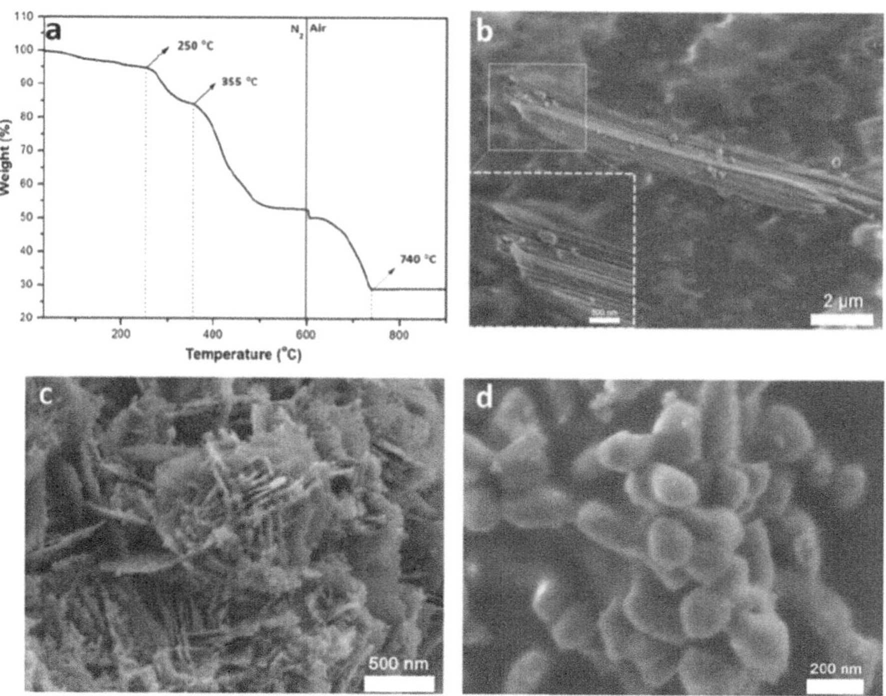

FIGURE 4.9 TGA curve (a) and SEM images of original MOF of LAC (b), carbon nanosheets after pyrolysis (c), CaO spheres after calcination (c). Reproduced with permission from Ref. [19].

medium, alkaline group-modified porous materials are used. Nitrogen-doped multiwall carbon nanotube is one such material that can be derived from the pyrolysis of MOF material with nitrogen-containing organic ligands. Shen and Bai studied the growth of multiwall carbon nanotubes by direct pyrolysis of MOF [13]. They heated $[Co(HTTG)(H_2O)_2]_n$ to 600°C for 1 h at a heating rate of 10°C/min and obtained nitrogen-doped multiwall carbon nanotube with the presence of cobalt nanoparticles around the tip. Tube length and diameter, amount, and type of nitrogen doping are highly dependent on the carbonization conditions. During pyrolysis, the cobalt present in the MOF decomposes and forms the Co nanoparticles which are surrounded by amorphous carbon originating from the organic linker molecule. At higher temperatures, carbon from the linker molecule gets dissolved at the surface of the Co nanoparticle and precipitates to form the graphitic tubular structure. During this process, the Co nanoparticle is raised to the tip of the growing carbon nanotube. The continued dissolution and precipitation of amorphous carbon lead to the growth of the carbon nanotube through the "tip-growth" mechanism as shown in Figure 4.10. Finally, the amorphous carbon, surrounding the Co nanoparticles that prevents their aggregation, is consumed and hence the Co nanoparticles will start aggregating to form a bulk cobalt substrate onto which the multiwall carbon nanotubes are eventually grown. Thus, once the amorphous carbon is exhausted, the final product of the

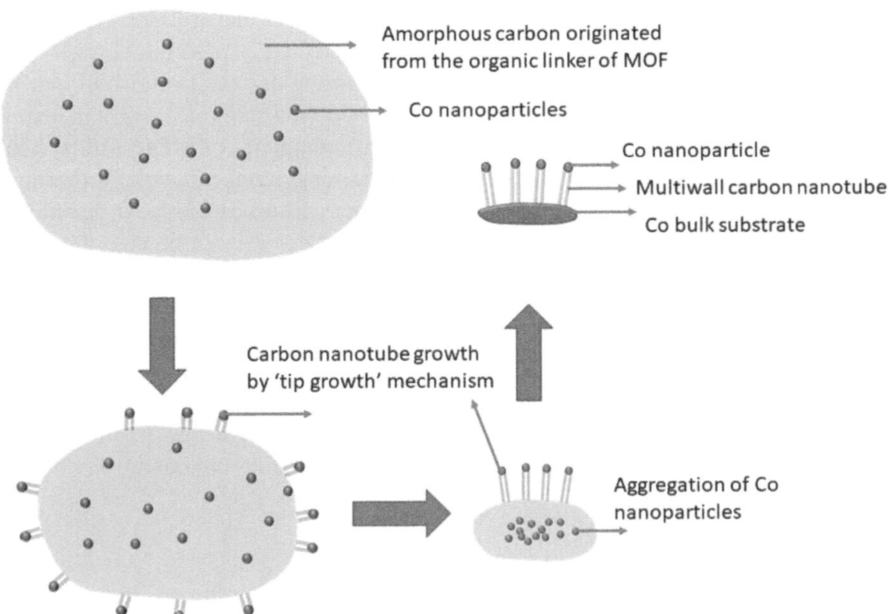

FIGURE 4.10 Schematic illustration of the proposed growth model for multiwall carbon nanotube.

multiwall carbon nanotube on the cobalt bulk substrate is obtained. The nitrogen content of multiwall carbon nanotube derived from $[Co(HTTG)(H_2O)_2]_n$ varied with the different carbonization conditions including the carbonization temperature and time. Carbon nanotubes obtained are closed from both ends by Co nanoparticles raised to the top and the cobalt bulk substrate at the bottom. This nitrogen-doped multiwall carbon nanotube was further applied for the selective removal of CO_2 over CH_4. Removal of CO_2 from natural gas is essential to enhance the efficiency of the natural gas and to prevent the CO_2-induced corrosion of natural gas pipelines. The nitrogen content of this multiwall carbon nanotube has a positive effect on the CO_2 selectivity over CH_4 thus exhibiting enhanced selectivity with increasing nitrogen content. The single-component isotherm corresponding to the adsorption of CO_2 fitted with the dual-site Langmuir-Freundlich adsorption model was further used to predict the multicomponent adsorption by ideal adsorbed solution theory. Though the selectivity decreases with increasing pressure beyond 8 bar, it shows an increase in selectivity which suggests that nitrogen-doped multiwall carbon nanotube is a promising candidate for the selective removal of CO_2 from the mixture of CO_2 and CH_4 which usually coexist in natural gas.

4.6.1.5 MOF-Derived Carbon Composite for CO_2 Adsorption

Adsorbents used for carbon capture are required to be hydrophobic in nature as during the carbon capture, water will also be adsorbed as both CO_2 and water have the same adsorption energy. Both water and carbon dioxide compete for the same adsorption sites which reduces the sites available for CO_2 adsorption. Therefore, it is essential to make the porous carbon adsorbent selectively adsorb CO_2 over water by making the surface hydrophobic. The porous carbon derived from MOF material itself is generally hydrophobic in nature. MOF-derived carbon has good textural properties derived from the parent MOF. The hydrophobicity of MDC will be better when the parent MOF material has a fluorine-containing organic ligand. Furthermore, to make the surface more hydrophobic the porous carbon can be post-fluorinated. Employing zinc-based MOF as a precursor to obtain the porous carbon has the advantage of low carbonization temperature because of the low evaporation temperature of zinc. Carbonization of zinc-based MOF (Zn-hfipbb) with perfluoro-containing organic linkers, (H2hfipbb = 4,4′-(hexafluoroisopropylidene) bis(benzoic acid)) and their application for CO_2 adsorption was studied at dry and humid conditions over N_2 adsorption [14]. This Zn perfluoro MOF-derived carbon composite micro- and mesopores obtained at a carbonization temperature of 700°C, demonstrated its potential as an adsorbent for CO_2 through physisorption even at humid conditions.

4.6.2 Hydrogen Storage

Concern over global warming witnessed more intense research on alternative fuels to conventional hydrocarbon-based fuel. Hydrogen is an alternate green fuel with high specific energy density. However, the onboard storage of hydrogen in the transport sector is challenging due to the low volumetric energy density, which limits its practical applications as a transportation fuel. Onboard hydrogen storage of 4–10 kg is essential to power a light vehicle to run for 300–500 miles. For this purpose, hydrogen can be stored in gaseous, liquid, and solid forms. The constraints in the storage of hydrogen in gaseous or liquid form limit its practical applications. For instance, storing hydrogen in the form of compressed gas requires a large volume of high-pressure tank so that enough hydrogen gas to fuel the vehicle can be stored. Similarly, to store hydrogen in liquid form requires cryogenic conditions due to its very low boiling temperature. The requirement of a high-pressure tank and cryogenic conditions are the major constraints for hydrogen storage in gaseous and liquid forms. Department of Energy, United States of America proposed short-term and long-term strategic pathways to overcome the challenges in realizing hydrogen as a transportation fuel. Though the immediate goal of hydrogen energy focuses on the cost reduction of compressed hydrogen gas storage using an appropriate high-pressure vessel, the long-term goal mainly focuses on cold or cryo-compressed hydrogen and material-based hydrogen storage. Materials in the form of sorbents, chemical hydrogen, or hydrides are explored for their hydrogen storage potential. Depending on the binding energy between the sorbent and hydrogen, different mechanisms can explain the hydrogen adsorption behaviour. There are four major mechanisms depicted in the Figure 4.11 that explain the interaction between hydrogen and sorbent surfaces based on the binding energy between hydrogen and sorbent materials.

(I)   Physisorption is the key mechanism of hydrogen adsorption on the surface of pure materials or materials with electronic configurations that do not support substantial adsorption. This non-dissociative molecular adsorption

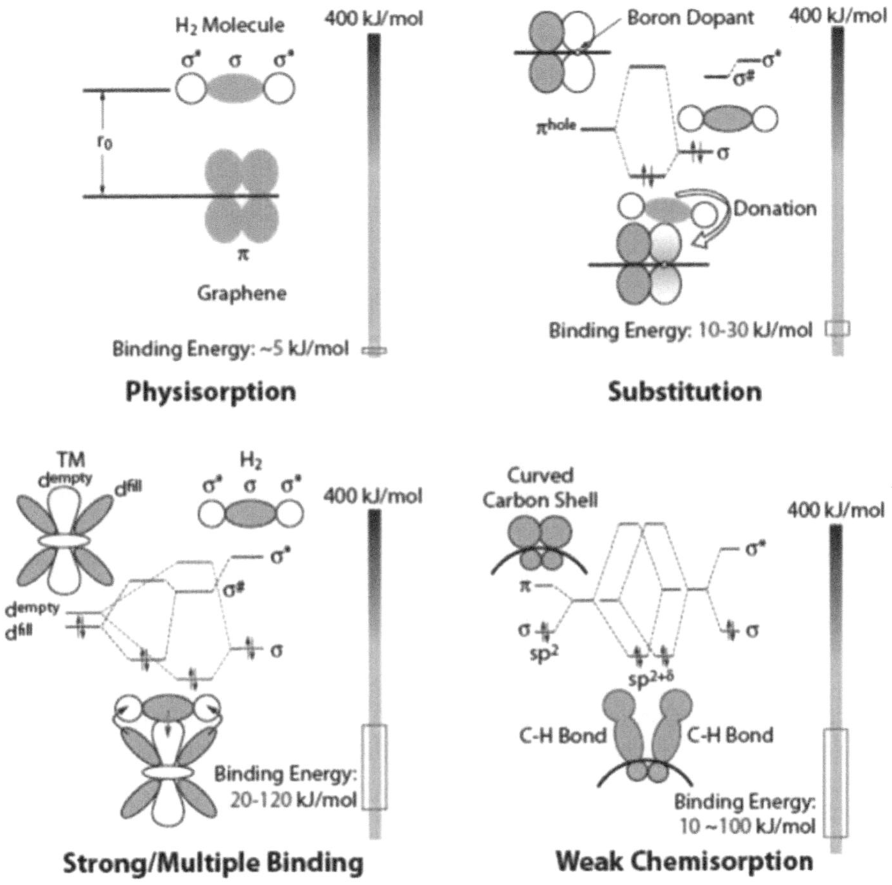

FIGURE 4.11 Mechanism of hydrogen storage on adsorbents. Reproduced with permission from Ref. [20].

of hydrogen happens through van der Waals interaction with the adsorption energy <5 kJ/mol. To enhance the adsorption capacity of the adsorbent through this mechanism, the sorbent material needs to be modified to access every hydrogen atom which requires enhanced specific surface area. This requires a porous structure with a pore size in the range of 0.7–1.5 nm which is sufficient to minimize the amount of open space.

(II) When the materials facilitate the electron transfer with hydrogen, it will lead to the formation of complexes. This enhanced dihydrogen binding will have binding energy in the range of 10–30 kJ/mol. To achieve the enhanced dihydrogen binding, the pure material has to be doped with suitable dopants to introduce the heterogeneity in the lattice structure which will enhance the electron interaction between hydrogen and sorbent.

(III) Multiple dihydrogen storage can happen with transition metal substrates through the electron transfer between the (i) σ orbital of hydrogen molecule and LUMO of transition metal and (ii) σ^* of hydrogen molecule and highest occupied molecular orbital (HOMO) of transition metal during which H-H bond gets stretched and results in atomic hydrogen adsorption on the sorbent surface. This multiple binding makes the bonding between hydrogen and the sorbent stronger with the binding energy ranging between 20 and 120 kJ/mol. In addition to the atomic hydrogen adsorption, this can accommodate hydrogen in molecular form through physisorption as well.

(IV) Another most commonly observed mechanism of adsorption of hydrogen on sorbent surface is through weak chemisorption with the binding energy of 10–100 kJ/mol [20].

4.6.2.1 Porous Carbon for H_2 Storage

Porous carbon being a lightweight material with a high specific surface area offers a solution to overcome the bottleneck of vehicular hydrogen storage. Researchers concentrate on enhancing the specific surface area of porous carbons through chemical or physical processes aiming to increase hydrogen uptake. Hydrogen storage through physisorption has the advantage of good recyclability due to its low energy requirement for adsorption and desorption. The materials with large pore volumes will have a large storage capacity for hydrogen through physisorption. In this regard, MOF being a porous material with a high specific surface area with an ordered microporous structure can be used for hydrogen storage. However, their poor aqueous and thermal stability restricts their intended application. The poor thermal stability of the MOF material causes the decomposition of MOF and subsequent carbonization easier. In contrast to the random pores of porous carbon obtained from alternate sources owing to challenges in controlling the pore structure, MOF precursors will result in well-defined and ordered pore structures similar to that of the parent MOF material. There are many MOF precursors reported to obtain porous carbon for H_2 storage applications. Carbonization of series of Zn-based MOFs ($Zn(BDC)TED_{0.5}$, where BDC-benzene dicarboxylic acid, TED-tetra ethylene diamine) modified with organic ligands with different functional groups namely –OH, NH_2, and CH_3, were reported for their hydrogen uptake capacity [21]. Among the different ligand-modified MOF-derived carbons, $–CH_3$ modification demonstrated the highest hydrogen storage capacity due to their favourable pore structure derived from the pore structure of the parent MOF. The BET surface area, pore volume, and nitrogen doping determine the hydrogen adsorption capacity. The hydrogen uptake capacity of carbon derived from three types of ZIFs with the same metal ion (Zn) but with different linker molecules with different aromaticity were studied by Wang et al. [6]. The microporous structure of CZIF8a, CZIF68a, and CZIF69a is evidenced by the Type-I adsorption isotherm with N_2 adsorption with the respective hydrogen uptake values of 2.59, 211, and 2.16 wt% at 1 atm, 77K. In spite of the lowest nitrogen content, CZIF8a with the highest BET surface area and pore volume shows the highest hydrogen uptake capacity. The presence of small-sized pores in the MDCs ensures the interaction between the MDC and hydrogen through physisorption which is reversible as well. The hydrogen uptake capacity of ZIF8-derived porous carbon is strongly influenced by carbonization temperature by altering the surface area and pore volume. Porous carbon obtained by carbonizing ZIF8, which serves as both template and precursor, along with the secondary precursor, furfuryl alcohol, at 1000°C exhibits a very high surface area of 3405 m^2/g with the corresponding pore volume of 2.58 cm^3/g compared to the carbonization temperature of 800°C which resulted in the surface area and pore volume of 2169 m^2/g and 1.50 cm^3/g, respectively. Type-I adsorption isotherm evidences the microporous characteristic of both C800 and C1000. The hydrogen uptake capacity at 1 atm and 77K corresponding to the carbonization temperature of 1000°C is significantly higher at 2.77 wt% compared to the value of 2.23 wt% obtained at the same conditions for a carbonization temperature of 800°C [22]. The effect of carbonization temperature on surface area and in turn the hydrogen uptake capacity can also be understood from the study performed on the carbonization of MOF-5 to obtain porous carbon [23].

Before carbonizing MOF-5, it was heated in the furfuryl alcohol atmosphere at a temperature of 150°C to enrich the carbon content by polymerizing the furfuryl inside the pores of MOF-5 which formed PFA/MOF-5. During the carbonization of PFA/MOF-5 at 530°C, MOF was decomposed and formed ZnO, which was subsequently removed by HCl washing. The enhanced specific surface area and pore volume presented in Table 4.6 achieved through HCl washing demonstrate that though the melting of Zn will facilitate the removal of Zn and resulting pore formation that renders an increased specific surface area and pore volume, the melting point of Zn is not essential to create pores in the carbon.

Increasing the specific surface area is one of the ways that will lead to a notable improvement in its capacity to adsorb and store H_2. The presence of hierarchical pores including macro-, meso-, and micropores will yield high specific surface area pore volume. Furthermore, enhanced interaction between H_2 molecules and the adsorbent surface stands out as a crucial approach to enhance the H_2 uptake capacity. When the diameter of the pores in the porous carbon reduces below 2 nm, the interaction between hydrogen and pore walls of carbon can be intensified. This intensified interaction between hydrogen and pore wall due to the presence of ultra micro pores, the pore size of which is in the order of kinetic diameter of hydrogen molecules, is highly beneficial for H_2 uptake. Efforts are underway to create adsorbents satisfying their multifaceted criteria including high specific surface area, high pore volume, the presence of hierarchical pores, and ultra microporosity in governing the performance of hydrogen uptake. Different types of isoreticular MOF materials (IRMOF-1, IRMOF-3, and IRMOF-8) were used as the template to produce porous carbon with hierarchical porous structure on direct carbonization [24]. Compared to the parent IRMOFs, the pore volume of different-sized pores including ultra micro-, meso-, and macropores, and the total pore volume increased in the case of MOF-derived carbon. Among the three MDCs, IRMOF-1-derived hierarchical porous carbon displayed an excellent hydrogen uptake capacity of 3.25 wt%. MDC derived from IRMOF-1 exhibited an enhanced total, ultra micro-, and meso-macropore volumes of 4.06, 0.63, and 3.05 cm^3/g, respectively, from the respective parent IFMOF-1 values of 1.45, 0.17, and 0.05 cm^3/g. While the existence of the hierarchical pores enhances the surface area and pore volume, and the ultra-micropore aids in enhancing the H_2-pore wall interaction, establishing mesopores that help in enhancing the surface area

TABLE 4.6
Illustrative MOF-derived materials for H$_2$ uptake

S.No	MOF-derived material	MOF precursor	Carbonization condition	Surface area (m²/g)	Pore volume	Ultramicropore volume	Micropore volume	H$_2$ uptake (wt%)	Ref
1	MDC-1	IRMOF-1	900°C,	3174	4.06	0.63	1.01	3.25	[24]
	MDC-3	IRMOR-3	5°C/min, 3 h	1678	2.01	0.49	0.66	2.1	
	MDC-8	IRMOF-8		1978	1.92	0.54	0.78	2.41	
								(at 77K, 1 bar)	
2	C800	ZIF-8 (commercial)	800°C, 8h	2181	1.5		0.9	2.23	[22]
	C1000		1000°C, 8h	3453	2.58		1.54	2.77	
								(at 77K, 1 atm)	
3	BF-900	ZIF-8 (commercial)	990°C, 3°C/min, 8 h	933	0.57		0.41	2.6	[25]
	ACBF-900			3188	1.94		1.16	6.2 (77K, 20 bar)	
4		MOF-5	1000°C, 5°C/min, 5 h	2393	1.13		0.34	2.7	[26]
			800°C, 5°C/min, 5 h	628	0.54		0.11	0.8	
								(77K, 1 bar)	
5	MOX-C	MOA	800°C, 5 h,	3770	2.62			2.98 (77K, 1 atm)	[27]
6		MOF-5	1000°C, 8 h	2872	2.06			2.6	[23]
7	NPC	[Zn$_3$(BTC)2.(H$_2$O)$_3$]$_n$	1000°C, 5°C/min, 5 h	1492	0.96		0.26	20.1 mg/g	[28]
								(77K, 1 bar)	
8	C800	IL@MIL-100(Al)	800°C, 7 h	2731	1.76			1.8	[8]
	NC800	MIL-100(Al) composited		2397	1.37			1.7	
	BNC800-2	with ionic liquids		1522	0.78			1.7	

still remains a challenge, especially without compromising the integrity of the ultra-microporous structure.

4.6.2.2 MOF-Derived Nanoscale Metal Hydride for H_2 Storage

Light-metal hydrides offer the advantages of high gravimetric and volumetric storage capacities which make them a potential candidate for solid-state onboard hydrogen storage material. However, the release of hydrogen from the metal hydride requires heating to a higher temperature due to the high H_2 desorption enthalpy arising from the strong chemical bond. This can be overcome by incorporation of the dopants, additives, alloying elements, or catalysts into the metal hydride. As an alternative, the metal hydrides can be brought down to nanoscale which reduces the H_2 desorption enthalpy relative to the bulk material and makes them a potential candidate for hydrogen storage. A porous template material with pores of less than 20 nm and a large pore volume is required to confine the metal hydride to the nanoscale and have large gravimetric and volumetric storage capacities. In recent years, in the realm of metal hydride hydrogen storage materials, researchers emphasized on confinement of metal hydride in porous materials including porous carbon, aerogel, and MOF with the primary objective of reducing the desorption enthalpy of hydrogen. This strategy lands them as a promising solid-state onboard hydrogen storage material. Nanoscale metal hydride derived from MOF material as a template enables the accomplishment of metal hydrides in the nanoscale. Nanoscale $NaAlH_4$ was obtained using HKUST-1 as the template which was infiltrated with a solution of $NaAlH_4$ in tetrahydrofuran and heated to 110°C under vacuum. The nanoscale $NaAlH_4$ obtained from the HKUST-1 template released 80% of hydrogen at 100°C lower than the bulk $NaAlH_4$ [29]. Magnesium hydride is a highly promising hydrogen storage material primarily attributed to its high theoretical hydrogen storage capacity of up to 7.6 wt% with excellent reversible properties [30–32]. Despite the high storage density and cost-effectiveness, magnesium hydride presents challenges due to its slow kinetics, the strong bond between Mg and hydrogen which necessitates high processing temperature, and its affinity towards oxygen during practical applications [33–35]. TMA-Ni MOF-derived Ni@C composited with MgH_2 by dispersing the former in the MgH_2 matric by ball milling provides the composite with better thermodynamics and kinetics demonstrated by the reduced adsorption enthalpy of hydrogen to −52.3 kJ/mol from −75.8 kJ/mol corresponding to pure MgH_2 [36]. Mg-Ni@C composite releases 2.5 wt% hydrogen at 573 K which continues to release 1.7 and 0.75 wt% hydrogen even at reduced temperatures of 523 and 498 K, respectively. The hydrogen storage capacity of MgH_2 was enhanced by bringing down the desorption enthalpy of hydrogen molecules by compositing MgH_2 with CoNi@C nanoparticles derived from flower-like CoNi-MOF through carbonization at 800°C with 2°C/min for 3 h [37]. On compositing Mg with Ni-Co@C, the formation of multiphase catalysts Mg_2Ni and Mg_2Co enhances the thermodynamics and kinetics of the Mg-based hydrogen storage system. The composite material of MgH_2, containing 10 wt% of CoNi@C, demonstrated a good hydrogen uptake of 6 wt% of H_2 reaching in 5 min. It achieved the saturated uptake of 6.4 wt% H_2 in just 20 min, displaying enhanced hydrogen desorption behaviour of 4.7 and 3.1 wt% H_2 released in 30 min at 573 K and 4.8 and 3.2 wt% H_2 released in 60 min.

4.7 CONCLUSION

MOF-derived materials have acclaimed global attention and made substantial strides in recent years, thanks to their properties stemming from the intriguing versatility in structural design of MOF, their tunable porosity, and remarkable surface area. In this chapter, we have highlighted the significance of MOF-derived materials in the context of the application of both doped and undoped MOF-derived porous carbon along with MOF-derived nanomaterials (including metal oxide nanoparticles and carbon nanotubes) towards adsorption and storage of CO_2 and H_2. We have outlined the diverse factors influencing the textural properties of MOF-derived materials and their impact on carbon capture and hydrogen storage. Despite the rapidly growing research progress in MOF-derived materials, there remain challenges to address in utilizing them as adsorbents for CO_2 and H_2 uptake. One notable area that requires attention is the need for more detailed mechanistic investigations. Enhancing the surface area of porous carbon is possible through the formation of hierarchical pores. However, creating mesopores to obtain the hierarchical pore structure still remains challenge due to the associated risk of compromising the structural integrity of ultra microporosity.

REFERENCES

1. Robens, E. Some Intriguing Items in the History of Adsorption. In *Studies in Surface Science and Catalysis*; Rouquerol, J., Rodríguez-Reinoso, F., Sing, K. S. W., Unger, K. K., Eds.; Elsevier, 1994; Vol. 87, pp 109–118. https://doi.org/https://doi.org/10.1016/S0167-2991(08)63070-0.
2. Deitz, V. R. comp. Bibliography of Solid Adsorbents--1943 to 1953. 1956, *566*.
3. Thommes, M.; Kaneko, K.; Neimark, A. V.; Olivier, J. P.; Rodriguez-Reinoso, F.; Rouquerol, J.; Sing, K. S. W. Physisorption of Gases, with Special Reference to the Evaluation of Surface Area and Pore Size Distribution (IUPAC Technical Report). *Pure Appl Chem*, 2015, *87* (9–10), 1051–1069. https://doi.org/10.1515/pac-2014-1117.
4. Pavlenko, V.; Khosravi H, S.; Żółtowska, S.; Haruna, A. B.; Zahid, M.; Mansurov, Z.; Supiyeva, Z.; Galal, A.; Ozoemena, K. I.; Abbas, Q.; et al. A Comprehensive Review of Template-Assisted Porous Carbons: Modern Preparation Methods and Advanced Applications. *Mater Sci Eng R: Rep*. Elsevier Ltd June 1, 2022. https://doi.org/10.1016/j.mser.2022.100682.
5. Srinivas, G.; Krungleviciute, V.; Guo, Z.-X.; Yildirim, T. Exceptional CO2 Capture in a Hierarchically Porous Carbon with Simultaneous High Surface Area and Pore Volume. *Energy Environ Sci*, 2014, *7* (1), 335–342. https://doi.org/10.1039/C3EE42918K.
6. Wang, Q.; Xia, W.; Guo, W.; An, L.; Xia, D.; Zou, R. Functional Zeolitic-Imidazolate-Framework-Templated Porous Carbon Materials for CO2 Capture and Enhanced Capacitors. *Chem Asian J*, 2013, *8*, 1879–1885. https://doi.org/10.1002/asia.201300147.
7. Bai, F.; Xia, Y.; Chen, B.; Su, H.; Zhu, Y. Preparation and Carbon Dioxide Uptake Capacity of N-Doped Porous Carbon Materials Derived from Direct Carbonization of Zeolitic Imidazolate Framework. *Carbon N Y*, 2014, *79*, 213–226. https://doi.org/https://doi.org/10.1016/j.carbon.2014.07.062.
8. Aijaz, A.; Akita, T.; Yang, H.; Xu, Q. From Ionic-Liquid@metal–Organic Framework Composites to Heteroatom-Decorated Large-Surface Area Carbons: Superior CO2 and H2 Uptake. *Chem Commun*, 2014, *50* (49), 6498–6501. https://doi.org/10.1039/C4CC02459A.

9. Gadipelli, S.; Guo, Z. X. Tuning of ZIF-Derived Carbon with High Activity, Nitrogen Functionality, and Yield – A Case for Superior CO2 Capture. *ChemSusChem*, 2015, *8* (12), 2123–2132. https://doi.org/10.1002/cssc.201403402.

10. Ma, X.; Li, L.; Chen, R.; Wang, C.; Li, H.; Wang, S. Heteroatom-Doped Nanoporous Carbon Derived from MOF-5 for CO2 Capture. *Appl Surf Sci*, 2018, *435*, 494–502. https://doi.org/10.1016/j.apsusc.2017.11.069.

11. Pan, Y.; Xue, M.; Chen, M.; Fang, Q.; Zhu, L.; Valtchev, V.; Qiu, S. ZIF-Derived in Situ Nitrogen Decorated Porous Carbons for CO2 Capture. *Inorg Chem Front*, 2016, *3* (9), 1112–1118. https://doi.org/10.1039/C6QI00158K.

12. Aijaz, A.; Fujiwara, N.; Xu, Q. From Metal–Organic Framework to Nitrogen-Decorated Nanoporous Carbons: High CO2 Uptake and Efficient Catalytic Oxygen Reduction. *J Am Chem Soc*, 2014, *136* (19), 6790–6793. https://doi.org/10.1021/ja5003907.

13. Shen, Y.; Bai, J. A New Kind CO2/CH4 Separation Material: Open Ended Nitrogen Doped Carbon Nanotubes Formed by Direct Pyrolysis of Metal Organic Frameworks. *Chem Commun*, 2010, *46* (8), 1308–1310. https://doi.org/10.1039/B913820J.

14. Park, J. M.; Lim, S.; Park, H.; Kim, D.; Cha, G.-Y.; Jo, D.; Cho, K. H.; Yoon, J. W.; Lee, S.-K.; Lee, U.-H. CO2 Capture Performance of Fluorinated Porous Carbon Composite Derived from a Zinc-Perfluoro Metal-Organic Framework. *Sep Purif Technol*, 2022, *302*, 121979. https://doi.org/10.1016/j.seppur.2022.121979.

15. Pan, Y.; Zhao, Y.; Mu, S.; Wang, Y.; Jiang, C.; Liu, Q.; Fang, Q.; Xue, M.; Qiu, S. Cation Exchanged MOF-Derived Nitrogen-Doped Porous Carbons for CO2 Capture and Supercapacitor Electrode Materials. *J Mater Chem A Mater*, 2017, *5* (20), 9544–9552. https://doi.org/10.1039/C7TA00162B.

16. Dunstan, T. M.; Donat, F.; Bork, H. A.; Grey, P. C.; Müller, R. C. CO2 Capture at Medium to High Temperature Using Solid Oxide-Based Sorbents: Fundamental Aspects, Mechanistic Insights, and Recent Advances. *Chem Rev*, 2021, *121* (20), 12681–12745. https://doi.org/10.1021/acs.chemrev.1c00100.

17. Armutlulu, A.; Naeem, M. A.; Liu, H.-J.; Kim, S. M.; Kierzkowska, A.; Fedorov, A.; Müller, C. R. Multishelled CaO Microspheres Stabilized by Atomic Layer Deposition of Al2O3 for Enhanced CO2 Capture Performance. *Adv Mater*, 2017, *29* (41), 1702896. https://doi.org/10.1002/adma.201702896.

18. Wang, Y.; Memon, M. Z.; Xie, Q.; Gao, Y.; Li, A.; Fu, W.; Wu, Z.; Dong, Y.; Ji, G. Study on CO2 Sorption Performance and Sorption Kinetics of Ce- and Zr-Doped CaO-Based Sorbents. *Carbon Capture Sci Technol*, 2022, *2*, 100033. https://doi.org/10.1016/j.ccst.2022.100033.

19. Liu, Z.; Lu, Y.; Wang, C.; Zhang, Y.; Jin, X.; Wu, J.; Wang, Y.; Zeng, J.; Yan, Z.; Sun, H.; et al. MOF-Derived Nano CaO for Highly Efficient CO2 Fast Adsorption. *Fuel*, 2023, *340*, 127476. https://doi.org/10.1016/j.fuel.2023.127476.

20. Simpson, L. *Hydrogen Sorptions Center of Excellence (HSCoE) Final Report*; US Department of Energy, 2005. www.energy.gov/sites/prod/files/2014/03/f12/hydrogen_sorption_coe_final_report.pdf

21. Li, R.; Han, X.; Liu, Q.; Qian, A.; Shen, H.; Liu, J.; Pu, X.; Xu, H.; Mu, B. Porous Carbon Materials with Improved Hydrogen Storage Capacity by Carbonizing Zn(BDC)TED0.5. *J Solid State Chem*, 2022, *314*, 123409. https://doi.org/10.1016/j.jssc.2022.123409.

22. Jiang, H.-L.; Liu, B.; Lan, Y.-Q.; Kuratani, K.; Akita, T.; Shioyama, H.; Zong, F.; Xu, Q. From Metal–Organic Framework to Nanoporous Carbon: Toward a Very High Surface Area and Hydrogen Uptake. *J Am Chem Soc*, 2011, *133* (31), 11854–11857. https://doi.org/10.1021/ja203184k.

23. Liu, B.; Shioyama, H.; Akita, T.; Xu, Q. Metal-Organic Framework as a Template for Porous Carbon Synthesis. *J Am Chem Soc*, 2008, *130* (16), 5390–5391. https://doi.org/10.1021/ja7106146.

24. Jae Yang, S.; Kim, T.; Hyuk Im, J.; Seung Kim, Y.; Lee, K.; Jung, H.; Rae Park, C. MOF-Derived Hierarchically Porous Carbon with Exceptional Porosity and Hydrogen Storage Capacity. *Chem Mater*, 2012, *24* (3), 464–470. https://doi.org/10.1021/cm202554j.

25. Almasoudi, A.; Mokaya, R. Preparation and Hydrogen Storage Capacity of Templated and Activated Carbons Nanocast from Commercially Available Zeolitic Imidazolate Framework. *J Mater Chem*, 2012, *22* (1), 146–152. https://doi.org/10.1039/C1JM13314D.

26. Segakweng, T.; Musyoka, N. M.; Ren, J.; Crouse, P.; Langmi, H. W. Comparison of MOF-5- and Cr-MOF-Derived Carbons for Hydrogen Storage Application. *Res Chem Intermed*, 2016, *42* (5), 4951–4961. https://doi.org/10.1007/s11164-015-2338-1.

27. Xia, W.; Qiu, B.; Xia, D.; Zou, R. Facile Preparation of Hierarchically Porous Carbons from Metal-Organic Gels and Their Application in Energy Storage. *Sci Rep*, **2013**, *3* (1), 1935. https://doi.org/10.1038/srep01935.

28. Wang, W.; Yuan, D. Mesoporous Carbon Originated from Non-Permanent Porous MOFs for Gas Storage and CO2/CH4 Separation. *Sci Rep*, 2014, *4* (1), 5711. https://doi.org/10.1038/srep05711.

29. Bhakta, K. R.; Herberg, L. J.; Jacobs, B.; Highley, A.; Behrens Jr., R.; Ockwig, W. N.; Greathouse, A. J.; Allendorf, D. M. Metal–Organic Frameworks as Templates for Nanoscale NaAlH4. *J Am Chem Soc*, 2009, *131* (37), 13198–13199. https://doi.org/10.1021/ja904431x.

30. Lutz, M.; Bhouri, M.; Linder, M.; Bürger, I. Adiabatic Magnesium Hydride System for Hydrogen Storage Based on Thermochemical Heat Storage: Numerical Analysis of the Dehydrogenation. *Appl Energy*, 2019, *236*, 1034–1048. https://doi.org/10.1016/j.apenergy.2018.12.038.

31. Sun, Y.; Shen, C.; Lai, Q.; Liu, W.; Wang, D.-W.; Aguey-Zinsou, K.-F. Tailoring Magnesium Based Materials for Hydrogen Storage through Synthesis: Current State of the Art. *Energy Storage Mater*, 2018, *10*, 168–198. https://doi.org/10.1016/j.ensm.2017.01.010.

32. Zhang, X.; Liu, Y.; Ren, Z.; Zhang, X.; Hu, J.; Huang, Z.; Lu, Y.; Gao, M.; Pan, H. Realizing 6.7 Wt% Reversible Storage of Hydrogen at Ambient Temperature with Non-Confined Ultrafine Magnesium Hydrides. *Energy Environ Sci*, 2021, *14* (4), 2302–2313. https://doi.org/10.1039/D0EE03160G.

33. Gremaud, R.; Broedersz, C. P.; Borgschulte, A.; van Setten, M. J.; Schreuders, H.; Slaman, M.; Dam, B.; Griessen, R. Hydrogenography of MgyNi1−yHx Gradient Thin Films: Interplay between the Thermodynamics and Kinetics of Hydrogenation. *Acta Mater*, 2010, *58* (2), 658–668. https://doi.org/https://doi.org/10.1016/j.actamat.2009.09.044.

34. Malahayati; Ismail; Mursal; Jalil, Z. The Use of Silicon Oxide Extracted from Rice Husk Ash as Catalyst in Magnesium Hydrides (MgH2) Prepared by Mechanical Alloying Method. *J Phys Conf Ser*, 2018, *1120* (1), 012061. https://doi.org/10.1088/1742-6596/1120/1/012061.

35. Rahwanto, A.; Jalil, Z.; Akhyar; Handoko, E. Desorption Properties of Mechanically Milled MgH2 with Double Catalysts Ni and SiC. *IOP Conf Ser Mater Sci Eng*, 2020, *931* (1), 012012. https://doi.org/10.1088/1757-899X/931/1/012012.

36. Yang, J.; Zhang, K.; Ma, Z.; Zhang, X.; Huang, T.; Panda, S.; Zou, J. Trimesic Acid-Ni Based Metal Organic Framework Derivative as an Effective Destabilizer to Improve Hydrogen Storage Properties of MgH2. *Int J Hydrogen Energy*, 2021, *46* (55), 28134–28143. https://doi.org/10.1016/j.ijhydene.2021.06.083.

37. Xing, X.; Liu, Y.; Zhang, Z.; Liu, T. Hierarchical Structure Carbon-Coated CoNi Nanocatalysts Derived from Flower-Like Bimetal MOFs: Enhancing the Hydrogen Storage Performance of MgH2 under Mild Conditions. *ACS Sustainable Chem Eng*, 2023, *11* (12), 4825–4837. https://doi.org/10.1021/acssuschemeng.2c07740.

5 Catalysis

5.1 INTRODUCTION

Catalysis is a process of accelerating the rate of a chemical reaction by reducing the activation energy required for that reaction to take place. It employs a non-consumable substance called a catalyst, which promotes the reaction to proceed in a different pathway that possesses a lower energy barrier to overcome and is much more energetically favorable than the uncatalyzed chemical reaction. This promotes the chemical reaction to occur in less time and makes it economical. Almost 80% of the chemical reactions on the industrial scale are driven by employing the catalyst which shows the potential of the catalytic-based reactions [1]. Figure 5.1 represents the pathways for a catalytic and non-catalytic process. As seen from the figure, the catalytic reaction requires a lower energy to cross the reaction barrier.

5.2 HISTORY OF CATALYSIS

The term "catalysis" was proposed by a Swedish chemist, Jons Jacob Berzelius, in the year 1835. It is derived from the Greek words' *kata* meaning "down" and *lyein* meaning "loosen". The early development of catalysis took place during the civilization period when man started to prepare alcohol by fermentation [3]. Valerius Cordus, in 1552, first time used an inorganic catalyst, sulphuric acid to convert alcohol to ether [4]. Parmentier, in 1781, studied that the reaction of potato starch with distilled water and potassium hydrogen tartrate gives more taste with the addition of acetic acid [5]. In 1813, Louis Jacques Thénard observed the decomposition of ammonia into nitrogen and hydrogen when it was allowed to pass over red-hot metals [3]. In the 18th century, many developments in catalysis were foreseen by eminent researchers. Some of them are Eilhard Mitscherlich who termed catalysis as a contact process, and Johann Dobereiner who studied the contact action. He also did potential research on the use of platinum and palladium iridium in the catalysis processes. He invented a lighter called the Dobereiner lamp consisting of hydrogen and platinum sponge. Wilhelm Ostwald, a Nobel awardee in 1909, investigated the chemical reactions that occur in the presence of acids and bases at finite rates. In the later stage, several developments were done by eminent researchers. Paul Sabatier

DOI: 10.1201/9781003432357-7

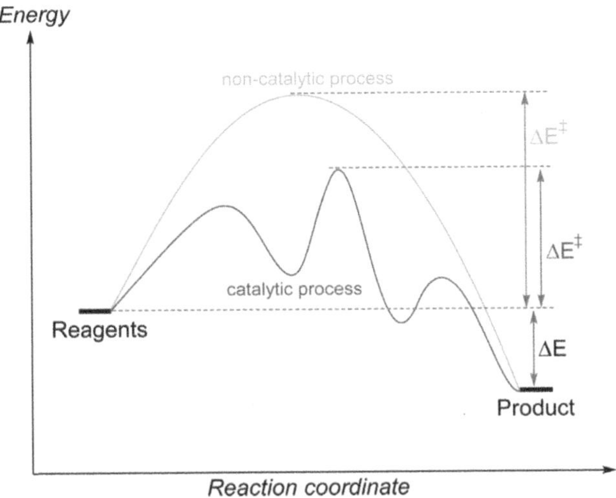

FIGURE 5.1 Representation of catalytic and non-catalytic reactions. Reproduced with permission from Ref. [2].

(Nobel awardee in 1912) and Jean-Baptiste Senderens discovered the phenomenon of catalytic hydrogenation to fix hydrogen based on metal catalysts. One of the most important catalysts that was discovered was the Pt-γ-Al$_2$O$_3$ which is being used to treat automobile exhaust gases through a selective catalytic reduction process.

5.3 CHARACTERISTIC FEATURES OF A CATALYST

The characteristic features of a catalyst are described as follows:

- A catalyst does not alter the chemical equilibrium which means the ratio of forward and backward reaction rates remains the same. Thus, there is no alteration in the equilibrium constant.
- A catalyst affects the kinetics however it does not have any effect on thermodynamics.
- A catalyst accelerates the pace of the forward reaction just as much as it does the reverse reaction.
- The effect of the catalyst can be seen only in kinetics but not in thermodynamics, i.e., if a chemical reaction is thermodynamically unfavorable, the catalyst cannot alter the conditions to make the reaction possible to happen.

5.4 CLASSIFICATION OF CATALYSTS

Catalysis can be classified into different types depending on the type of catalytic substance employed and the nature of the catalytic reaction [6]. Some of the important classifications of catalysis are:

Based on the nature of the catalyst:

a. Solid phase catalyst
b. Liquid phase catalyst
c. Gas phase catalyst

Based on the starting material from which the catalyst is prepared:

a. Organic (enzymes, organic acids, etc.)
b. Inorganic (metal, metal oxides, inorganic acids, bases, etc.)

Based on the phase of the catalyst concerning the reactants:

a. Homogeneous catalysis: The catalyst and the reactants, in a Homogeneous catalysis exist in the same phase (most often liquid phase). The catalyst dissolves in the medium along with the reactants to lower the activation energy required for a chemical reaction.
b. Heterogeneous catalysis: The catalyst and the reactants in a Heterogeneous catalysis are in different phases. Catalysts can take the form of solids which can interact with the reactants that are either in liquid or gaseous phase.

Based on the catalytic action:

a. Acid-base catalysis: In this process, the addition of acid or a base which acts as a proton donor or acceptor catalyzes the chemical reaction without the consumption of acid or base.
b. Photocatalysis: This process employs a photocatalyst that can be excited in the presence of light to interact with the chemical species present in the reaction.
c. Electrocatalysis: It is one of the classes of heterogeneous catalytic processes. In electrocatalysis, the catalyst can take the form of the electrode itself or can be present at the electrode surface which participates in the electrochemical reaction and increases the rate of reaction by transfer of electrons to or from the reactant species.
d. Enzyme catalysis: Enzymes such as proteins are used as catalysts, especially in biochemical reactions that happen at a specific site on the catalyst called an active site.

Based on the state of the catalyst:

a. Nanoclusters: In this type, a cluster or group of atoms are dispersed on a catalyst surface which can drive the reaction pathway.
b. Nanoparticles: Here, nanoparticles are dispersed on the catalyst surface which can drive a chemical reaction with lower activation energy.
c. Single-atom catalysis: In this type of catalysis, unlike a cluster of atoms, a single atom is dispersed on a catalyst surface to give rise to a highly selective active site which can drive the chemical reaction to a lower activation energy.

The selection of a suitable catalyst plays a crucial part in deciding whether the reaction takes place or not as suggested by the French chemist Paul Sabatier. If the interactions between the reactants and the catalyst are too weak, then there is no possibility of chemical reaction to occur. Such a situation is described as catalyst poisoning due to reactants. If the interactions between them are too strong, then the dissociation of products from the catalyst surface is difficult. This is termed as catalyst poisoning due to products.

5.5 METAL–ORGANIC FRAMEWORKS AS CATALYSTS

Catalysts can take any form starting from atoms, and molecules to large structures. Earlier many materials were explored for their potential use as catalysts such as zeolites, graphene, and carbon-based materials. Recently, metal–organic frameworks (MOFs), an intriguing family of porous crystalline compounds, have attracted scientific enthusiasts due to their profound characteristic features and properties. These reticular materials offer a wide range of opportunities for catalytic applications due to the presence of high-densified catalytic sites, large pore openings that can accommodate the transport of reactants, enormous surface areas, etc. The possibility of post-synthetic modification of MOFs by incorporating functional groups that can enhance the catalytic activity contributes to preferring MOFs over other materials. The abundant unsaturated metal Lewis acid sites present in MOFs and the Lewis basic organic linkers serve as the active catalytic sites in MOFs. Besides these advantages offered by MOFs, they possess instability issues due to moisture, fragile bonding between the metal centre and the organic linker, and poor thermal stability. However, these disadvantages can be overcome by converting them into much more stable porous materials than the parent MOFs without compromising on the inherent porosity and surface area. These materials are termed MOF-derived materials.

5.6 MOF-DERIVED MATERIALS AS CATALYSTS

MOF-derived materials are formed from pristine MOFs through several synthesis routes such as direct pyrolysis of MOFs, incorporating guest molecules coupled with pyrolysis, MOFs embedded with various substrates followed by pyrolysis, and solution infiltration. The parent MOFs serve as the sacrificial template or precursor to give rise to various ordered structures such as porous graphitic structures, N-rich carbon-based materials, metal/metal oxide doped structures, and heteroatom doped porous structures. The composition of the MOF, i.e., metal centres (e.g., Cu, Ni, Fe, Co, etc.) and the organic linker (e.g., C, H, O, S, N) can be retained in the derived materials which are crucial in catalysis. In addition to this, they possess a higher surface area, higher porosity, tunable morphology, and uniform heteroatom doping than their pristine MOFs. These features offer significant contributions in catalytic applications such as hydrogen evolution reaction (HER), oxygen evolution reaction (OER), water splitting, CO_2 reduction or conversion, water-gas shift reactions, hydrogenation reactions, and so on. MOF-derived materials show greater advantages than traditional porous materials not only in terms of surface area and porosity but also in modulating the structures of derived materials without any additional template or precursor. Due to the reticular chemistry and adjustable morphology possessed by

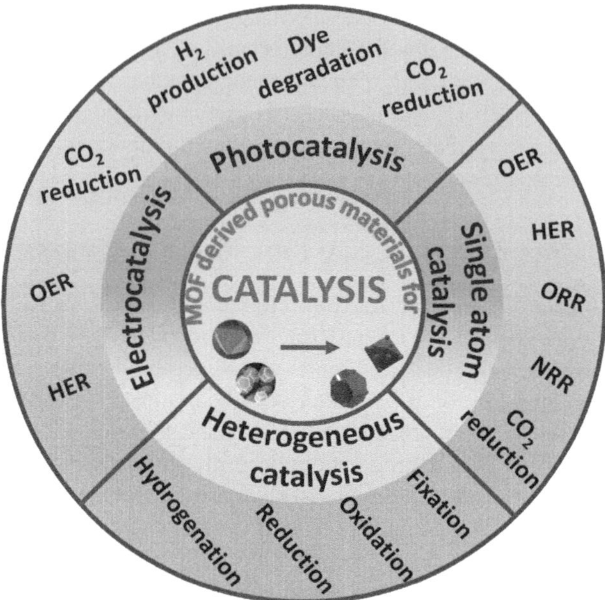

FIGURE 5.2 Schematic representation of various applications of MOF-derived materials.

MOFs, they help to control the structure, alter the electronic properties of catalysts, and provide easy accessibility to the active sites. Hence diversified porous structures can be developed from MOFs. The structure-property-performance relationship elucidates that the diversity of these derived materials with abundant active sites and electronic properties can be adjusted by the pre-design of MOFs that are well suited to a particular application. The higher degree of crystallinity present in MOFs leads to the formation of ordered porous structures while maintaining structure integrity. The metal species will be well dispersed throughout the porous framework and the aggregation of the metal species can be prevented during pyrolysis due to the well-ordered MOF precursor. Having these advantages and possessing enhanced features when compared to parent MOFs, MOF-derived substances are being explored in various applications such as electrocatalysis, photocatalysis, heterogeneous catalysis, and single-atom catalysis. Herein, the applications of MOF-derived materials for various catalytic applications are discussed. A schematic representation of various applications of these materials is shown in Figure 5.2.

5.7 MOF-DERIVED MATERIALS FOR ELECTROCATALYTIC REACTIONS

5.7.1 Electrocatalysis

Electrocatalysis is a process employing a substance called electrocatalyst to aid electrochemical reactions. It is a chemical reaction that happens at the interface of an electrode and an electrolyte solution and is assisted or accelerated by this electrocatalyst. The reaction involves the transport of electrons at the electrode-electrolyte interface. Electrocatalysis is a basic process in electrochemistry that is widely used in numerous applications such as OER, HER, and CO_2 reduction reactions (CO_2RR). Concerning electrocatalysts, pristine MOFs possess disadvantages such as low electrical conductivity and low stability which hinder electrochemical performance. To overcome these challenges, MOF-derived materials were explored as catalysts for various electrocatalytic reactions. In the present chapter, the applications of MOF-derived materials for various applications are discussed.

5.7.2 MOF-Derived Materials for Oxygen Evolution Reaction (OER)

OER is a basic electrochemical process that happens in water-splitting reactions, fuel cells, metal-air batteries, etc. where O_2 is evolved electrochemically from H_2O. OER plays a vital role in many energy storage and conversion processes. The governing equation for the OER process is shown below.

Alkaline medium: $4\ OH^- \rightarrow O_2 + 2H_2O + 4e^-$
Acidic medium: $2H_2O \rightarrow O_2 + 4H^+ + 4e^-$

The reaction involves the oxidation of H_2O molecules at the anode which releases O_2, four protons, and four electrons. To drive this reaction, a significant amount of energy is required as input. In reality, an excess amount of potential is required than the standard potential (1.23 V) required for water splitting, termed as overpotential. Generally, this overpotential is considered as the potential to achieve a current density of 10 mA/cm^2 (E_{10}). The material with low overpotential requirements can be regarded as the best electrocatalyst for OER reaction. The other parameters required to evaluate the OER activity are exchange current density, Tafel slope, turnover frequency, faradaic efficiency (FE), stability, etc. Hence, it is required to find suitable electrocatalysts for OER with low overpotential, excellent OER activity, long-term stability, and cyclability. Various kinds of MOF-derived materials were explored as catalysts for OER reactions which are explained below.

For an efficient electrocatalytic reaction to take place, the catalyst should aid rapid mass and electron transport and provide abundant catalytic sites for the surface reactions. In this regard, MOF-derived carbonaceous materials with tunable porosity and enormous electrical conductivity act as promising materials for OER reactions. They possess several advantages such as

1. Structural diversity, which can provide a huge amount of active sites and ensure fast mass transport.
2. Homogeneous distribution of metal centres in carbon-based materials can improve catalytic activity.
3. Formation of redox-active sites via hetero atom doping.
4) Facilitating the incorporation of guest species inside the pores of carbon materials that can provide various functionalities.

Two types of MOF-derived carbon materials were studied for the OER activity, metal-free and metal-containing carbon materials. Metal-free carbon materials have poor stability and lower OER activity due to the low graphitization degree that occurs during pyrolysis. Hence, not many studies were done on metal-free carbon materials for OER activity. However, the doping of heteroatoms into metal-free carbon structures which can assist electrocatalytic activity was explored for OER. Especially, doping of pyridinic N and C=O into the carbon structure can enhance the OER performance [7]. Zn-based MOFs were explored for OER since the Zn metal can evaporate during pyrolysis. An N-doped carbon synthesized by direct carbonization of ZIF-8 which had profound pyridinic N and C=O was evaluated for OER activity [8]. The best OER activity was achieved with an overpotential of 1.71 V in 0.1 M KOH electrolyte. In another instance, N and B were co-doped into the carbon framework which showed a low onset potential of 1.38 V in 0.1M KOH but it showed an overpotential of about 1.8 V. The overpotential decreased to 1.55 V when the electrolytic medium was changed to 6 M KOH, but there was a decline in current density [9]. These studies enabled us to understand the importance of heteroatom doping while stressing on improving the stability to achieve exemplary OER activity.

When it comes to metal constituting carbonaceous materials, they tend to exhibit excellent OER activity due to their stability under OER potentials. Metal-based carbon materials occupy the majority of share in the list of OER electrocatalysts. Another advantage of the presence of metal in the framework is that the surface oxidation during the reaction will comparatively improve the OER activity. The metal can take many forms inside the framework like metal nanoparticles, metal oxides and hydroxides, metal chalcogenides, nitrides, phosphides, etc. They are formed with the aid of externally added reducing agents during pyrolysis or the addition of precursor materials for these metal sources during pyrolysis. Controlling the metal content in MOF precursor and suitably selecting the metal precursor source affects the final product after pyrolysis. It can be achieved by regulating various parameters during pyrolysis such as temperature and time. Ai et al synthesized an onion-like graphene layer wrapped with Ni nanoparticles derived from direct pyrolysis of Ni-MOF which exhibited excellent OER performance with an onset potential of 1.53 V and a higher current density [10]. The overpotential (E_{10}) was found to be only 370 mV which is just behind the traditional precious metal-based OER catalyst RuO_2 (300 mV). MOF-derived bimetallic materials have shown enhanced OER activity in comparison to their monometallic counterparts. Wang et al used a Fe adsorbed-Ni MOF as a precursor which was annealed in the presence of hexamethylenetetramine (HMTA) to get carbon-coated NiFe alloy nanoparticles [11]. In this, the arrangement of the Ni, Fe alloy nanoparticles, i.e., the crystal structure was regulated by varying the annealing temperature. The HCP structural arrangement showed a greater OER activity than the FCC arrangement with E_{10} of 226 mV and a Tafel slope of 41 mV/dec. In another instance, Nitrogen-doped carbon loaded with Ni/Fe nanoparticles was prepared by direct pyrolysis of Ni/Fe polyphthalocyanine. The resultant material showed excellent performance with a Tafel slope of 78 mV/dec.

Apart from designing electrocatalysts with metal NP, metal oxides, hydroxides, and composites were also investigated for OER activity. These architectures have

FIGURE 5.3 SEM-EDS mapping of three different morphologies obtained after pyrolysis of ZIF-67 at various conditions (a, b) Typical rhombic dodecahedron (d, e) Flower kind, and (g, h) Hollow sphere. Reproduced with permission from Ref. [13].

shown promising results due to their distinct morphology, diversified functionalities, and controllable structures. Optimization of several synthesis parameters is crucial to get the desired morphology and controllable features. These desired structures could be obtained by direct pyrolysis or by post-synthetic incorporation of guest species into the framework of a MOF-derived porous carbon matrix. Ma and his group designed an efficient electrocatalyst based on Co-MOF [12]. The MOF was grown in Cu foil to get nano-wired morphology. A similar morphology was retained even after pyrolysis resulting in Co_3O_4/C nano wire structure. For a catalyst to work efficiently towards the OER process, morphology plays a crucial part. It was proved in many instances that the optimized morphology exhibited the best catalytic OER performance. For instance, Yang et al. utilized ZIF-67 to derive porous composites with different morphologies. Besides preserving the regular dodecahedron morphology, two other morphologies, hollow sphere and flower kind of morphological features were obtained after pyrolysis as shown in Figure 5.3. Interestingly, the one with flower morphology exposed the best OER output [13]. This reinforces the fundamental concept of material science – the "Structure-property-performance" relationship.

5.7.3 MOF-Derived Materials for Hydrogen Evolution Reaction (HER)

HER is an electrochemical process in which hydrogen is evolved during water-splitting reactions. It is similar to the OER process except that hydrogen species get adsorbed in the catalytic centres of the catalyst during the reaction. HER reaction is an alternate renewable way of producing H_2 by splitting the water and generating the energy by burning the released H_2 in the presence of O_2 serving as a fuel source. The criteria for determining a good electrocatalyst for HER is its hydrogen adsorption energy. Similar to the OER process, the governing parameters defined to quantify the HER activity include onset potential, Tafel slope, etc. The reactions involved in HER process are shown below:

Alkaline medium: $2H_2O + 2e^- \rightarrow H_2 + 2OH^-$
Acidic medium: $2H^+ + 2e^- \rightarrow H_2$

It is a two-electron transfer process where hydrogen is produced at the cathode. Like the OER process, a good HER catalyst should have low overpotential, low cost, high selectivity, and performance. MOF-derived materials offer a wide variety of advantages as catalysts for the HER process such as ultra-high porosity, dense active catalytic sites, excellent mass transport, and charge transport. These make them stand out as a promising candidates replacing the existing high-cost Pt-based catalysts. Many transition metal-based MOF-derived materials were explored for HER activity. Hou and his co-workers reported a Co containing N-doped porous carbon matrix derived from ZI-67/GO material [14]. The resultant catalyst exhibited an overpotential of 229 mV at a current density of 10 mA/cm² in an acidic medium. Ni-MOF was pyrolyzed by Xu and his group to obtain Ni NP incorporated in a graphene structure which showed better HER activity with a low overpotential of 205 mV at a current density of 10 mA/cm² in basic medium [15]. The high performance can be attributed to the presence of rich metal centres that act as catalytic sites for HER. Moreover, the presence of graphene structure exhibited high conductivity and good mass transport took place due to its highly porous nature.

Apart from incorporating NP, their oxides, sulphides, and composites were also incorporated into the carbon structure which made them outperform many existing catalysts. Tian et al. synthesized a Co_9S_8/MoS_2 doped carbon catalyst derived from Co-MOF which showed HER activity with an overpotential of 233 and 194 mV in 0.5M H_2SO_4 and 1M KOH, respectively, at a current density of 10 mA/cm². The presence of carbon matrix prevented the NP from agglomeration and corrosion by anchoring them to its defective sites and suppressing them from oxidation [16]. Zou et al. prepared a CoP embedded in BCN nanotubes from the pyrolysis and phosphating process of Co-MOF. Due to the hollow structure of CoP@BCN nanotubes, the material showed an excellent HER activity with an overpotential of 87 and 215 mV in acidic and basic mediums, respectively [17]. Taking the advantage of structure-performance relationship, Pan et al. prepared a flower-like Co-N-C composite that achieved remarkable catalytic activity with excellent stability over a long duration of time [18].

Since corrosion is a problem in many electrochemical processes, developing corrosion-resistant materials is of high importance. Carbides of W and Mo are resistant to corrosion. Hence the introduction of these compounds into the porous substrate will improve the corrosion resistance as well as electrochemical activity. Wu and his co-workers reported a novel material introducing MoC_x species into cavities of NENU-5 with subsequent pyrolysis [19]. A synergistic effect was established between the nano carbide particles and the porous matrix. It was observed that as the particle size of carbides started decreasing, the catalytic activity increased.

5.7.4 MOF-Derived Materials for Electrocatalytic CO_2 Reduction

The electrocatalytic CO_2 reduction reaction (ECRR) is a critical step in the realms of sustainable energy conversion and chemical synthesis. It entails the use of electrocatalysts to facilitate the electrochemical conversion of carbon dioxide (CO_2) into value-added products, therefore reducing greenhouse gas emissions and using renewable energy resources. The primary goal of ECRR is to avoid the thermodynamic problems associated with CO_2 reduction by making the multi-electron transfer steps required for the creation of suitable reaction intermediates easier to accomplish. The key steps involved in the CO_2 reduction process are:

1. The adsorption of the CO_2 flux on the catalytic active sites.
2. Surface reactions between the catalyst and the CO_2 molecules to produce intermediates with 1 electron – 1 proton transfer process.
3. The desorption of the products.

Depending upon the interaction between the CO_2 molecules and the catalytic active centres, reaction pathways, and the experimental conditions, a wide variety of products are obtained starting from C_1 (constituting one carbon) to C_3 and many more.

MOF-derived materials were explored for ECRR due to their ultra-high porosity, high density active metal catalytic centres. They offer excellent mass transport with the pores of the framework [20]. The type of product formed depends upon the metal active centre present in the derived material. For instance, ZIF-based derived materials were commonly seen to form CO products while Cu-MOF-derived materials were observed to form multi-carbon containing products [21–23]. Among various metals explored for CO_2 reduction, Cu has shown excellent reactivity over long periods [24, 25]. However, the application was limited by its selectivity towards the formation of various species. The production of hydrocarbon from CO_2 using a Cu electrode generally requires negative potential depending upon the product formed [26]. But thermodynamic calculations revealed that +0.17 V vs RHE is required. The need for such large overpotential was explained by Peterson et al in a detailed mechanism using DFT calculations about the ability of Cu to reduce CO_2 to various alcohols [27]. To improve the selectivity, Cu species were supported on other catalyst surfaces with a highly porous and conductive nature. Nam et al synthesized an HKUST-1-based Cu/carbon structure that could selectively reduce CO_2 to various carbon products. The methodology involved was to control the formation of the Cu clusters that could shift the CO_2 electroreduction to generate various products. The selectivity was achieved by regulating the coordination number of Cu-Cu dimer species in the Cu clusters [28]. In another work, the carbonization of HKUST-1 MOF yielded alcoholic products with FE of 45.2%–71.2% at −0.1 to −0.7 V RHE [29]. The Cu clusters were distributed uniformly in the porous carbon framework which was responsible for the selective reduction of CO_2 to ethanol. Ethanol was formed with a high yield at a lower overpotential of 190 mV. DFT calculations were performed to identify the reaction pathways as shown in Figure 5.4.

The mechanism involves the transfer of one electron to CO_2 to form CO_2^{*-} which is then reacted with a proton-electron pair to form $^*COOH$. In the subsequent step, the $^*COOH$ reacted with one more proton-electron pair to form formic acid. The formation of ethanol and methanol took place by the reduction of $^*COOH$ species in the subsequent steps in another reaction pathway with processes involving dihydroxylation and hydrogenation. Due to the uniform distribution of Cu species on the highly porous conductive carbon framework, the enhancement in the ethanol and methanol formation was observed.

Jonathan et al synthesized various kinds of Cu-based MOF-derived porous carbon materials for ECRR of CO_2 into ethanol and methanol using gas diffusion electrodes [22]. The distribution of Cu clusters on the porous framework was responsible for the selective conversion of CO_2 into alcohols. The resultant materials showed excellent performance with good FE and stability.

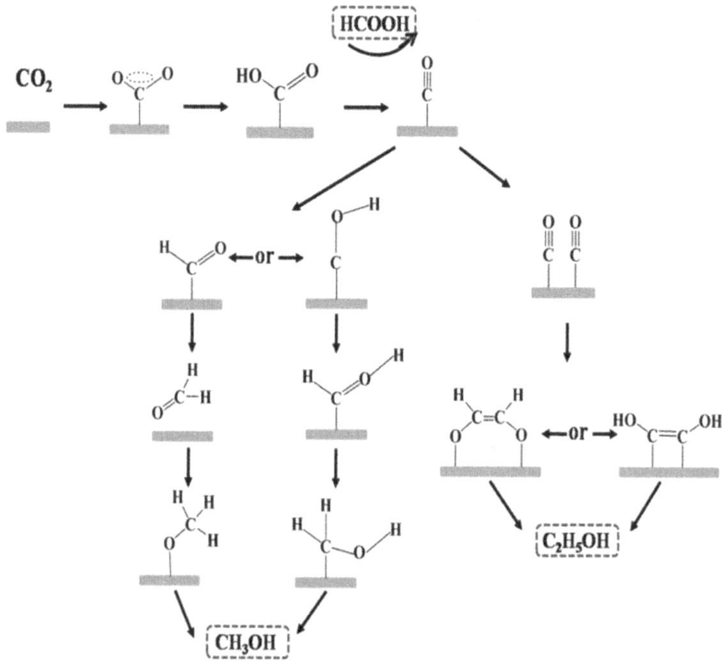

FIGURE 5.4 Proposed reaction mechanism of ECRR of CO_2 to formic acid, ethanol, and methanol on oxide-derived Cu/carbon catalyst. Reproduced with permission from Ref. [29].

Selective formation of a product is a crucial step in the reduction of CO_2. The selectivity depends upon how strongly the species are bonded to the metal catalyst. New strategies are being developed to transform CO_2 into a single carbon product. Strategies such as altering the oxidation states of Cu species distributed inside the carbon structure [30] and surface area [31] will help in the selectivity of particular species. Recently, it was shown that the introduction of strains in the lattice could lead to various phenomena causing selective formation of products [32, 33]. In this regard, Kaili et al synthesized a $Cu@Cu_xO$ core-shell framework derived from the calcination of HKUST-1 for selective conversion of CO_2 to C_2H_4. The calcination of HKUST-1 at 265°C led to a mixed phase of CuO and Cu_2O in the structure which caused a lattice tensile strain [31]. This was responsible for high CO_2 activation. Under the application of potential, the oxidation states started varying and promoted the C_2H_4 conversion and regulating the CH_4 formation. The resultant MOF-derived material showed a FE of 51% towards C_2H_4 and about 70% for C_{2+} species. Similar work was carried out by Junyu et al where the Cu/Cu_2O nanoparticles with variable oxidation states were prepared for electrocatalytic reduction of CO_2 [34]. High current densities of 25.15 mA/cm^2 were obtained at an applied potential of 0.79 V vs. RHE. In another work, the structure of the catalyst was modified by incorporating Cu_2O/Cu nanospheres in an N-doped porous carbon structure derived from HKUST-1 [35]. The catalyst showed enhanced activity and selectivity towards formate with a low overpotential of 380 mV. The doping of N species in the framework led to the formation of *OCHO species which caused selective generation of formate product with FE of 70.5% at −0.68 V vs RHE. These studies suggested that the selective formation of a product during CO_2 reduction can be regulated by altering the porosity of the structure, and composition, tailoring the oxidation states, and engineering the crystal lattice structure.

When it comes to ZIF-based derived porous structures for ECRR, many studies have been extensively explored. Zn-based ZIF materials were studied due to the abundant availability of Zn source and also due to its selective formation of CO [36]. Research revealed that the ECRR using Zn-based catalysts will depend upon the structure and morphology. Won et al synthesized a hierarchical Zn catalyst that exhibited different behavior when the morphology was changed. The Zn (101) facet showed the best performance in the generation of CO whereas the Zn (002) facet exhibited HER activity [37]. A similar feature was observed in Ag-MOF-derived three-dimensional Ag dendrites with numerous edge sites that served as active catalytic centres and selective reduction was observed at the (111) facet because of its high exposure and weak binding energy with CO [38]. From these studies, it could be concluded that selectivity could be enhanced by incorporating active defective sites in, a 3D porous network that can enable sufficient mass transfer and electron transport. Therefore, tailoring these basic features of the material could help in achieving greater heights in reducing CO_2 to obtain various products. Taking this into consideration, Yuhan et al prepared a porous ZnO material derived from ZIF-8 with huge grain boundaries that acted as the catalytically active centres for the selective reduction of CO_2 to CO [39]. The performance of as-derived material was greater than Zn foil and ZnO with rod morphology with an FE of 86.7% and a current density of 16.1 mA/cm^2 at −1.2 V vs RHE. Other kinds of strategies were explored such as dispersing the highly active single-atom catalytic centres in carbon matrix which have low metal coordination numbers using various synthetic approaches. This way of modifying the local structure is an elegant approach to expose highly active centres for selective reduction of CO_2. In a facile approach, Zhao et al exchanged the Zn-ZIF-8 with Ni metal nodes using post synthetic solvent exchange and then pyrolyzed to get N-doped carbon with highly exposed Ni centres [40]. This rational design gave an excellent CO formation with FE of 71.9% and, the current density of 10.48 mA/cm^2 with 0.89 V overpotential. A similar strategy was followed to control the coordination number by pyrolysis temperature to achieve atomically dispersed Co centres on the matrix of N-doped carbon for selective reduction of CO_2 to CO [41]. Changing the pyrolysis temperature led to a change in the coordination number of Co-N centres. The best performance was shown by the one with coordination number two with FE of 94% and an overpotential of 520 mV. The current density was seen to be 18.1 mA/cm^2 with a record turnover frequency of 18200 h^{-1}. Apart from Co-N sites, Fe-N sites were also observed to show enhanced selective reduction of CO_2 [42]. To have good performance, a controlled synthesis procedure needs to be carried out to deposit these sites only on the surface of the calcined product. If these sites are present deeply inside the matrix, they cannot be accessed easily to show catalytic activity [43]. The Fe-N sites were dispersed on the porous matrix using ammonium ferric citrate as an iron source through a post-synthetic modification approach.

5.8 MOF-DERIVED MATERIALS FOR PHOTOCATALYSIS

The term photocatalysis refers to the catalytic process that takes place in the presence of light. A photocatalyst is a material that drives a chemical reaction when it is exposed to light. Photocatalysis involves the generation of electron-hole pairs when the photocatalyst is irradiated with light energy. All the photocatalysts usually are semiconductors. They possess electrical conductivity at ambient temperature in the presence of light. Generally, semiconductors are materials with a band gap (E_g) lying between 1.5 and 3 eV. When a photocatalyst is irradiated with a certain wavelength, the photonic energy is absorbed by the electron present in the valence band and gets excited to the conduction band leaving behind a hole in the valence band. Thus, upon excitation with a certain wavelength, an electron-hole pair is generated. The electron is used to reduce the acceptor species, and the hole is utilized to oxidize the door substances. Thus, the photocatalysis process involves the creation of both reduction and oxidation phenomena simultaneously. The effectiveness of this process depends on the band gap of the photocatalyst and the way it interacts with the light in a particular wavelength and also the quick transfer of electrons and holes [44, 45]. Photocatalysis plays a prominent role in the degradation of dyes, contaminants, H_2 generation, antifouling, energy storage, water splitting reactions, demineralization, CO_2 conversion, etc.

Two types of photocatalytic processes include:

1. Homogeneous catalysis: Process in which both the catalyst and the guest species are in one phase, i.e., gas, liquid, or solid.
2. Heterogeneous catalysis: Process in which the catalyst and the guest species are in different phases.

Semiconducting metal oxides such as TiO_2, ZnO, CuO, and WO_3 were explored as photocatalysts. However, they face certain limitations such as wide band gap, incomplete conversion of organic matter, and electron-hole recombination [46–48]. To overcome these limitations, MOFs were explored as sacrificial templates upon pyrolysis to build porous structures made of oxides and sulphides which can retain the pristine structures of MOF, high porosity, and geometry so that excellent mass transfer and electron-hole transport take place. MOF-derived materials have been explored as photocatalysts in recent years for a wide range of applications. The versatility of these materials lies in the presence of ultra-high porosity facilitating an excellent interaction between the charge carriers and the substrate material thereby hampering the electron-hole recombination process and facilitating the pollutant transport. Moreover, the pyrolysis of MOF leads to the formation of metal sulphides and oxides, a few of which are semiconductors. Besides, due to the highly porous nature, rapid electron transport is established which is necessary for photocatalysis. In the present chapter, the application of MOF-derived materials for various photocatalytic applications are discussed.

5.8.1 Photocatalytic Degradation of Dyes

Due to the contamination of water resources through industrial waste, the availability of potable freshwater resources is diminishing day by day. Industrial waste generally consists of dyes, heavy metals, toxic chemical waste, oil spills, etc. These pose a great threat to human health as well as aquatic life. Hence the need to address these issues is of great importance. Photocatalytic degradation has been extensively utilized to convert these pollutants into non-toxic substances.

Porous metal oxide-based photocatalysts can be derived from the direct calcination of MOF by retaining the structure of pristine MOF. C-doped ZnO/TiO_2 was prepared in this method using Zn/Ti MOF with varying molar ratios of Zn to Ti for photocatalytic degradation of Rhodamine B [49]. The best activity was shown by the material which was calcined at 600°C and outperformed the activity shown by pure ZnO and TiO_2. The enhancement can be attributed to the C-content present in the material which was responsible for rapid charge transfer by trapping the electrons.

Metal sulphides such as CdS, Sb_2S_3, and In_2S_3 were greatly explored as photocatalysts due to their abundance and availability at a cheaper cost. They exhibit good conductivity and sensitivity in the presence of light. In_2S_3 with various morphologies such as nanorods, nanosheets, and microspheres showed enhanced photocatalytic activity and attracted researchers for the removal of toxic chemicals present in wastewater [50–53]. The availability of these sulphides in various morphologies can prevent catalyst aggregation during the degradation process [54, 55]. In this regard, Yuan et al synthesized MOF-derived In_2S_3 nanorods from MIL-68 precursor for the degradation of methyl orange (MO) [56]. The results indicated that the degradation process was governed by the superoxide radicals ($\cdot O_2^-$) and holes (h^+). The photocatalyst showed a 97% degradation rate in 20 min.

ZnO showed promising results when compared to TiO_2 because of its excellent chemical stability, utilization of light to a greater extent, and low cost. Several porous structures based on ZnO exhibits improved catalytic activity when precursors such as

MOF-5 and ZIF-8 are used. Additional agents such as non-metals [57] and metals [58] were doped with this to improve the utilization of visible light range and chemical stability. CuO was utilized to broaden the light utility and reduce electron-hole recombination in combination with ZnO and was proved to be an excellent photocatalyst [59]. At the same time, modifying the morphology of the CuO/ZnO structures and adopting various synthesis routes gave excellent results in the photodegradation of dyes and organic pollutants [60–62].

5.8.2 Photocatalytic H_2 Production

As the need for switching to renewable energy sources is increasing day by day, the route for producing clean and renewable energy must be explored. In this regard, photocatalysis plays a crucial role in water-splitting reactions to produce hydrogen (H_2) as a fuel source for many potential applications. For this purpose, the photocatalytic reactions must satisfy at least some of the prerequisites below.

1. Excellent light-capturing capability
2. Electron-hole pair generation
3. Good charge separation and transfer
4. Redox reactions

Many semiconducting nanomaterials such as TiO_2 [63], g-C_3N_4 [64], ZnO [65], and CdS [66] were explored for the H_2 generation, but are limited to lower efficiencies by their agglomeration characteristics, low energy trapping capability, electron-hole recombination, etc. Many alternatives were proposed to overcome these disadvantages. Recently, the incorporation of a co-catalyst on the surface of a photocatalyst was explored, including the loading with noble metals [67] and transition metal sulphides (TMS) [68]. However, due to their cost, doping with noble metals did not make a good impression. Providing hollow structures to the photocatalyst offers good mass transport and charge carrier efficiency. MOF serves as a templates by producing hollow framework structures through various synthesis routes. Because of its coordination environment, MOF-derived materials prevent the loss of light energy and trap a greater amount of incident light by multiple refraction and scattering mechanisms [69].

The formation of heterojunctions by combining two different semiconductors was found to be an elegant way of enhancing photoactivity. Zhou et al. synthesized $CoCeO_x$/g-C_3N_4 nanocomposite heterojunction catalyst for H_2 production from Co-doped Ce-MOF with an H_2 formation rate of 1050 μmol/g/h [70]. Due to the presence of Ce and Co, there was an improvement in the light absorption. Moreover, because of the spherical/sheet morphology of the composite, more active sites were generated which promoted charge separation and transportation. A similar kind of research was established by Su et al. to synthesize ZIF-67-derived Co_3O_4/$NiCo_2O_4$ heterojunction catalyst with double shell polyhedron morphology [69]. Because of this morphology, the transport distance for the charge carriers was reduced and abundant active centres were created which suppressed the electron-hole recombination. Similar ZIF-67 was utilized to make Co_3O_4 containing N-doped carbon $(Co_3O_4@C)$ which served as a co-catalyst with TiO_2 that gave a H_2 formation rate of 11400 μmol/g/h [71]. $Co_3O_4@C$ was responsible for the suppression of the recombination process and also it improved the separation efficiency of the generated charge carriers. The resultant composite catalyst outperformed the catalytic activity of pure TiO_2 which was, in turn, low due to the faster charge recombination phenomenon. In another work, Wang et al. used CuS, a low-cost and better performing co-catalyst to replace the precious noble metals [72]. CuS was loaded on CeO_2/ZnS composite derived from ZIF-8 MOF precursor which gave a superior activity in H_2 production. The addition of a co-catalyst resulted in the formation of a Schottky barrier thus enhancing the H_2 generation efficiency 13470 µmol/g/h. Other works with respect to the H_2 production through MOF-derived materials are shown in Table 5.1.

5.8.3 PHOTOCATALYTIC CO_2 REDUCTION

Similar to electrocatalytic CO_2 reduction, photocatalytic reduction plays an important role in converting CO_2 into many useful products. In order to break the C=O bonds present in CO_2 molecules and convert them into useful products, a significant amount of energy is required [73]. Since the availability of solar energy is abundant, light-driven CO_2 conversion is an economical and renewable process of reduction of CO_2 [74]. The reduction process proceeds with multiple reactions, each giving a specific product with corresponding standard reduction potential. The photocatalytic CO_2 reduction/conversion mainly involves three steps.

1. Trapping or harvesting solar energy in a broad range of spectrum
2. Photogenerated charge separation and transportation
3. Conversion of adsorbed CO_2 at the active sites

Many photo-responsive metal oxide semiconductors such as TiO_2, ZnO, In_2O_3, and Co_3O_4 [75–77] were explored because of their light harvesting nature. However, because of their low efficiency in trapping light and only in a small range of spectrum, they were unable to selectively convert CO_2 into valuable products with high yield and efficiency. Besides, the CO_2 photoreduction reaction requires the adsorption of CO_2 molecules onto the photocatalyst surface. In this regard, MOF-derived materials that can inherit the characteristic features of pristine MOFs can serve as excellent photocatalysts for the conversion of CO_2.

Ye and his co-workers synthesized Fe-NP containing carbon matrix nanocomposite derived from MIL-101 (Fe) through simple pyrolysis [78]. The resultant material showed an excellent catalytic activity by combining the advantages of plasmon resonance features of Fe NP and the porous structure of the carbon matrix which resulted in selective conversion to CO. The material gave a CO output of 2196.17 µmol after exposure to solar energy for 120 min. The surface plasmon resonance phenomenon and the optoelectronic properties of NP left a significant impact on the photoconversion of CO_2. Carbonaceous materials resulting from the pyrolysis of various MOFs gave interesting results in the catalytic conversion of CO_2. The pyrolyzed material consisted of metal active sites in the matrix of graphitic carbon which possessed superior electrical conductivity. A similar kind of work was carried out by Khaletskaya et al. where Ag NP was incorporated into the NH_2-MIL-125(Ti) for the conversion of CO_2 to CH_4 [79].

The required morphology and properties of the catalyst were obtained by the controlled pyrolysis of the MOF precursor. The addition of Ag NP was distributed uniformly over the framework thereby significantly enhancing the CO_2 conversion ability. These works showed the ability of incorporation of NP in improving the photocatalytic activity of MOF-derived materials for selective conversion of CO_2 to various organic products.

Recently, the creation of heterojunctions in the photocatalysts was seen with an escalating interest since they exhibited excellent charge separation and transport and enhanced photocatalytic activity. Many reports were available where heterojunctions were created with various morphologies such as nanotubes, nanorods, and hollow structures that promoted selective conversion of CO_2 to various organic fuels [80–83]. Cheng and his co-workers had grown ZIF-67 on InOF-1 nanorod morphological structures followed by acid-etching and low-temperature heat treatment to give a hollow structured Co_3O_4/In_2O_3 bimetallic oxide framework which gave a high CO output of 4828 ± 570 μmol/h/g over six runs. The Co centres served as active sites and In_2O_3 centres acted as CO_2 carriers promoting a high yield of CO [84]. Sulphides such as $CuInS_2$, CdS, NiS_x, MnS, and In_2S_3 possess rich surface defects and they are semiconductors. Several sulphide-based heterojunction photocatalysts were prepared to improve the catalytic activity towards the reduction of CO_2. The usual construction of sulphide-based semiconductor heterojunction involves the synthesis of individual sulphide structures and then assembling them to give a framework. MOFs promote a facile route for the synthesis of these heterostructures through the pyrolysis process giving rise to sulphides as well as inheriting the porous nature of pristine MOF structure thus facilitating excellent mass transport and diffusion. For instance, Tan et al prepared MnS/In_2S_3 p-n heterojunction structures derived from Mn-incorporated MIL-68(In) for photocatalytic reduction of CO_2 [85].

5.9 MOF-DERIVED MATERIALS FOR HETEROGENEOUS CATALYSIS

Heterogeneous catalysis is a prominent branch of catalysis in which the catalyst occurs in a different phase than the reactants (solid, liquid, or gas). This sort of catalysis is important in many industrial processes and has multiple uses in the production of chemicals, fuels, and materials, degradation, and reduction of pollutants in water and air. Unlike homogeneous catalysis, in which the catalyst and reactants are in the same phase, heterogeneous catalysis entails a separate phase separation. Solid catalysts are often utilized in the presence of gaseous or liquid reactants.

Three types of heterogeneous catalytic reactions can be observed:

1. **Surface type catalysis**: It is the basic heterogeneous catalytic process where the reactions occur on the pore walls and the outer surface of the catalyst. The rate of reaction is directly proportional to the surface area of the catalyst.
2. **Bulk type(I) catalysis**: In this type of catalysis, the reactants are absorbed inside the interpolyanion space of the crystal structure, and the solid catalyst behaves as a solution. Hence it is termed as pseudo-liquid catalysis.

3. **Bulk type (II) catalysis**: In this process, even though the governing catalytic reaction occurs on the surface, the whole solid catalyst is involved in the redox reaction where rapid transport of protons and electrons takes place. The rate of reaction is directly proportional to the volume of the catalyst.

Heterogeneous catalytic reactions include a broad variety of chemical transformations that take place at the interface between a solid catalyst and reactant molecules in various phases (gas, liquid, or solid). Some examples of frequent heterogeneous catalytic reactions are oxidation, hydrogenation, dehydrogenation, Fisher-Tropsch synthesis, selective catalytic reduction (SCR), etc.

MOF-derived materials possess unique advantages due to the presence of active metal clusters in their frameworks that act as catalytic sites. The catalytic active centres present in pristine MOFs can be inherited to their derived materials while maintaining the framework topology and stability. Many porous materials derived from MOF were explored for heterogeneous catalysis such as porous carbon, metal/metal-oxide, metal/metal-sulphides, and other composite structures applied in various catalytic organic reactions. They exhibit not only excellent performance but also facilitate excellent mass transport in various catalytic organic reactions. In the present context, a brief idea about MOF-derived materials for various heterogeneous catalytic reactions are provided.

Liquid phase organo-catalysis:

a. Oxidation reaction
b. Reduction reactions
c. CO_2 fixation reactions

Gas-phase catalysis:

a. CO oxidation reaction
b. Fisher-Tropsch` process
c. CO_2 hydrogenation reaction

5.9.1 MOF-Derived Materials for Liquid Phase Organo-Catalysis

5.9.1.1 Oxidation Reactions

MOF-derived materials were applied in various oxidation reactions such as oxidizing alcohols, hydrocarbons, and oxidative coupling of amines. Oxidation reactions of these substances result in the formation of organic compounds that can serve as starting materials for many organic reactions or can be directly utilized. For example, the oxidation of primary alcohols to aldehydes, secondary alcohols to ketones, and aliphatic alcohols to carbonyl compounds was achieved through MOF-derived compounds which demonstrated excellent selectivity and high yield [86]. Besides these materials can also be applied to reactions such as epoxidation of styrene [87], oxidation of cis-cyclooctene to epoxy cyclooctane [88], oxidation of hydrocarbons such as ethylbenzene, cyclohexane, toluene, oxidation of amines such as benzylamines

TABLE 5.1
MOF-derived materials for various heterogeneous oxidation processes

S.No	Catalyst	MOF precursor	Substrate	Reaction	Ref.
1	Cu@C	Cu-BTC	Benzyl alcohol	Alcohol oxidation	[90]
2	Co@C-N (800)	ZIF-67	Methanol & benzyl alcohol	Alcohol oxidation	[91]
3	Pd/ZDC@ mesoSiO$_2$	Pd^{2+}/ZIF-8@ mesoSiO$_2$	Benzyl alcohol	Alcohol oxidation	[92]
4	Co-CoO@C-N	ZIF-67	Styrene	Alcohol esterification	[93]
5	ZnS-N/C	Zn(bpp)(Hsba)	Cis-cyclooctene	Epoxidation reaction	[88]
6	C-N-Co	ZIF-67	Styrene	Styrene epoxidation	[87]
7	Co-CoO@C-N	ZIF-67	Styrene	Styrene epoxidation	[94]
8	N-doped TiO$_2$@N-C	NH$_2$-MIL-125(Ti)	Benzylamine	Amine oxidation	[95]
9	Co@C-N600	Co-MOF	Aldehydes	Amine oxidation	[96]
10	V-N-C-600	NH$_2$-MIL-101(V)	Benzylamine	Amine oxidation	[97]
11	Cr$_3$C$_2$/rGO	MOF-101-GO	Ethylbenzene	Hydrocarbon oxidation	[98]
12	Ni@C-N	Ni-MOF	Alkanes	Hydrocarbon oxidation	[99]

to secondary imines [89]. All these reactions were governed by various factors such as the structure-property relationship, composition, incorporation of secondary guest species into the framework, and pyrolysis temperature.

Non-noble and noble metal incorporated carbon or N-doped carbon obtained from pyrolysis of MOFs were proven to act as excellent catalysts, especially for the oxidation of alcohols. Metal containing N-doped carbon porous catalysts was found to play a significant role in epoxidation reactions. Regarding the oxidation of amines and hydrocarbons, metal and metal free heteroatom incorporated carbon structures provided profound catalytic active sites. Some of the applications of MOF-derived porous materials for various oxidation reactions are listed in Table 5.1.

5.9.1.2   Reduction Reactions

Reduction reactions play a vital role in the production of fine chemicals with a high order of purity, fuels, and pharmaceuticals. These include the production of alcohols from various substituted hydrocarbons, such as aldehydes, ketones, and esters; the reduction of nitriles and many other nitro compounds into amines; and hydrogenation reactions of ketones, aldehydes, and nitro compounds. Noble metal and non-noble metal-incorporated porous carbon structures derived from various kinds of MOFs were studied for the reduction of many compounds. Since the cost of precious metals such as Pd, Pt, and Au, etc. is high, the focus is shifted to transition metal-based catalysts. The porous structures will stabilize these metal species facilitating the catalytic activity. Selectivity towards a particular species is of high importance in these reactions. To achieve this, various synthesis conditions were adopted to obtain the preferred morphology and the composition of the final framework.

So far, Co, Ni-based MOFs were utilized to prepare carbon or N-doped carbon, especially for hydrogenation reactions. For the first time, Li et al. derived a Co@C-N catalyst from Co-MOF for the hydrogenation of acetophenone to 1-phenylethanol [100]. The resultant catalyst showed a selectivity of 98% and a conversion efficiency of 72%. The presence of basic sites in the catalyst promoted the transfer of protons to the metal surface thereby generating the metal-hydride compounds which further participated in the reaction. Multicomponent porous substrates with two different metals such as Ni and Co were studied for hydrogenation reactions of benzonitrile [101] and phenol [102]. This led to the formation of abundant active sites for hydrogenation with an improved conversion efficiency of about >98% and high selectivity. Recently, a novel work was reported utilizing a mixed-linker strategy to incorporate Co NPs and Co single atoms in the framework of a graphitic shell derived from Co-DABCO-TPA@C-800. Because of the presence of dual linkers, high N content was dispersed in the structure which facilitated the formation of various amines [103]. The coupling reactions were carried out using ketones and aldehydes in combination with nitro compounds, amines, and ammonia to get compounds from primary to tertiary amines.

Other types of combinations such as metal oxide incorporated porous structures, metal NP/metal oxide structures, and metal-free-based frameworks were synthesized for heterogeneous reduction reactions. For example, Murugesan and his co-workers reported a facile strategy to incorporate Co NP and Co_3O_4 into graphene shells for hydrogenation of aliphatic, aromatic, and heterocyclic nitriles to primary amines [104]. Jiang et al. synthesized two different materials based on metal oxide incorporation, γ-Fe_2O_3 from Fe-MIL-88A [105] and Co-CoO@NC [106] from ZIF-67 for hydrogenation of various compounds to amines. Van Nguyen et al. reported a metal-free B, N co-doped porous carbon framework for catalytic reduction of 4-nitrophenol from boron substituted ZIF-8 material [107]. The resultant catalyst showed excellent reduction capability with a lower activation barrier. This was due to the synergistic effect among the B, N, and O atoms in the catalyst framework creating more dense active sites for catalysis. A similar strategy was also carried out by Liu and his co-workers for the reduction of nitro compounds [108]. Other types of MOF-derived materials for various reduction reactions are summarized in Table 5.2.

TABLE 5.2
MOF-derived materials for various heterogeneous reduction processes

S.No	Catalyst	MOF precursor	Substrate	Ref.
1	CeO_2-Co_3O_4	CeO_2-ZIF-67	NO	[109]
2	Pd-MDPC	ZIF-67	Phenylacetylene	[110]
3	Ni@C/Ni-MOF-74	Ni-MOF-74	1-Octene	[111]
4	PCN-224-T	PCN-224	4-Nitrophenol	[112]
5	Co@Pd/NC	ZIF-67	Nitroarenes	[113]
6	PtCo@NHPC-1	ZIF-8	Nitro compounds	[114]
7	Pt_3Co/PCT	NH2-MIL-125 (Ti)	Cinnamaldehyde	[115]

5.9.1.3 CO_2 Fixation Reaction

As the rising levels of CO_2 in the atmosphere creates an alarming situation for the global community, it is important to curb this impact and protect the environment. Among the various techniques available for tackling the CO_2 issue, CO_2 fixation is one of the prominent game-changers that attracted scientific attention. The abundant volume of CO_2 can be converted into several value-added products such as polycarbonates. ZIF-based derivatives showed excellent CO_2 fixation in combination with epoxides to produce cyclic carbonates. The presence of N content along with metal species acted as catalytic sites. The mechanism first follows the adsorption of CO_2 in the porous framework, subsequent carbamate formation, and conversion to cyclic carbonates in the presence of epoxides. Ding et al. synthesized ZnO incorporated N-doped carbon material obtained by pyrolysis of ZIF-8, consisting of distinct active sites such as ZnO, hydroxyl, carboxyl species, pyridinic N content which showed enhanced selectivity and conversion of CO_2 to cyclic carbonates [116]. In another work, Toyao and his group developed a series of N-doped carbon materials through various ZIF structures [117]. Among all the structures obtained, the one derived from ZIF-9 consisting of partially oxidized Co NP exhibited better catalytic activity due to the presence of dense active catalytic Lewis's acid and basic sites.

5.9.2 MOF-Derived Materials for Gas Phase Reactions

5.9.2.1 CO Oxidation

Carbon monoxide (CO) is one of the significant environmental pollutants whose volumes have been rising for decades. Hence, fruitful utilization of CO by converting it into an alternative product seemed to be an acceptable strategy. Earlier, many noble metal species such as Pd, Pt, and Au and transition metal-based catalysts were utilized for the oxidation of CO. Recently, MOF-derived materials gained escalating interest due to their unique characteristic features. Especially, Cu-based MOF-derived porous structures in the form of metal oxides and its composites showed good impact in the conversion of CO. Lu et al. synthesized CuO/Cu_2O porous structures from pyrolysis of Cu-BTC which showed good CO conversion due to its huge surface area [118]. Materials derived from Cu-BTC with various morphologies such as octahedral, cubic, nanorod, and nanowires were obtained under controlled synthesis conditions. The conversion capabilities obtained were different for different morphologies which proved the structure-performance dependence relationship. Composites were prepared by loading oxides such as CeO_2, TiO_2 along with CuO/Cu_2O which showed excellent activity under mild experimental conditions [119–121]. Other types of porous-derived materials derived from Cu^{2+}@Ce-UiO-66 showed a selectivity of 99.5% towards CO oxidation in a H_2-rich gas flow [122]. A novel material Co/C-600 obtained from ZIF-67 showed the capability of oxidizing CO even at $-30°C$ besides retaining the moisture stability [123].

5.9.2.2 Fischer-Tropsch Synthesis

The Fischer-Tropsch synthesis (FTS) is a chemical procedure employed for transforming carbon monoxide (CO) and hydrogen (H_2) gases (syngas) into hydrocarbons, particularly lengthy alkanes. It is used to manufacture synthetic fuels like artificial gasoline, diesel, and other hydrocarbon-based products using diverse carbon-rich source materials such as coal, natural gas, and biomass. This process has been utilized in fuel and chemical production since the early 20th century, signifying its historical significance. To address the problems such as the huge deactivation rate faced by traditional materials, MOF-derived porous substances were found to be potential candidates for the FTS process. Many Fe, Co-based derived materials from ZIF, MOF-74, etc. frameworks with high active sites seemed to be good catalysts due to their cost-effectiveness. Santos and his group synthesized Fe NP encapsulated in a carbon framework with different Fe/C contents [124]. The as-prepared catalyst exhibited a conversion efficiency of up to 77%. In another work, Co-derived porous structures based on ZIF-67, and Co-MOF-74 were obtained for the FTS process [125]. Because of the larger pore size of Co@C obtained from pyrolysis of Co-MOF-74 than that of ZIF-67, it exhibited excellent activity with respect to the FTS process. Because of the larger pore size, the diffusion of the target species became easy, and efficient mass transport took place. Zhang and his co-workers prepared a novel Co@C catalyst from MOF which showed excellent FTS efficiency of about 83% with no indication of deactivation of the catalyst [126]. This could be attributed to the porous carbon shell type of structure that prevented the agglomeration of Co NP inside the framework.

5.9.2.3 CO_2 Hydrogenation

To achieve net zero carbon, CO_2 hydrogenation, also called as reverse water-gas shift reaction (RWGS), is one of the promising technologies. This reaction will reduce CO_2 to produce CO. When compared to noble metal catalysts such as Pt, Pd, and Rh, MOF-derived porous materials offer several advantages such as dense active sites, selectivity, and better conversion efficiency. Cu-based catalysts exhibited promising results, but they suffered from aggregation problems when the temperature went beyond 300°C. Hence it was inserted into the porous carbon matrix by pyrolysis of Cu-MOF. The resultant framework prevented the aggregation and showed excellent catalytic performance. For instance, Zhang and his co-workers prepared Cu@C and Cu/Zn@C from their respective precursor MOFs and achieved selectivity of 100% towards CO even at a higher temperature of 500°C [127]. Gandara Loe et al. pyrolyzed MIL-100(Fe) to get three different kinds of catalysts with in-situ incorporation of Rh. These catalysts showed tremendous catalytic activity below 500°C [128]. In another work, CeO_2 was impregnated into the Cu-MOF and then pyrolyzed to give CuO_x/CeO_2 for the hydrogenation of CO_2. The addition of CeO_2 promoted the RWGS reaction due to its multiple oxidation states [129].

5.10 MOF-DERIVED MATERIALS FOR SINGLE-ATOM CATALYSIS

5.10.1 SINGLE ATOM CATALYSIS

Single-atom catalysis is a new and exciting field of catalysis, which plays an important role in chemical reactions and industrial processes. While standard catalysts are usually made up of nanoparticles or bigger structures, single-atom (SA) catalysis deviates from the norm by using individual atoms as catalysts. This subject has received a lot of interest because of its potential to improve catalytic efficiency and selectivity while reducing dependence on precious or rare metals in catalytic applications. Metal atoms or other catalytic components are placed singly onto a supporting substance, such as a metal oxide or a carbon-based material, in the realm of single-atom catalysis. These single atoms play the function of active sites, stimulating chemical reactions. What distinguishes single-atom catalysts is their ability to produce remarkable catalytic efficiency and selectivity. Each atom acts as an independent catalytic centre, providing precise control of the catalytic process while limiting unwanted side reactions. Single-atom catalysis is used in a wide range of chemical processes, including the catalytic conversion of molecules such as carbon monoxide (CO), carbon dioxide (CO_2), and nitrogen (N_2). It also includes processes related to energy storage, such as the hydrogen evolution reaction and the oxygen reduction reaction, as well as practical applications in sectors such as environmental cleanup. The selection of support materials is crucial in single-atom catalysis. The choice of support materials may have a significant impact on the stability and reactivity of the individual atoms serving as catalysts. Metal oxides, carbon-based materials such as graphene, and zeolites are common support materials, with each having a unique influence on the performance of single-atom catalysts. MOFs can serve as precursors to obtain single-atom catalysts (SAC) with single metal atom sites dispersed uniformly on the porous framework exhibiting profound performance. These are indeed applied in various electrocatalytic applications and the synthesis of organic compounds.

5.10.2 MOF-Derived SAC for Electrocatalytic Applications

5.10.2.1 Oxygen Reduction Reaction (ORR)

For SAC, ORR might proceed in two reaction pathways as shown below.

$$O_2 + 2H + 2e^- \rightarrow H_2O_2; \; E = 0.70 \text{ V}$$
$$O_2 + 4H + 4e^- \rightarrow 2H_2O; \; E = 1.23 \text{ V}$$

The particle size of the SAC catalysts plays a vital role in the activity of ORR due to the differences in adsorption-desorption energies of various intermediates. When the NPs are brought down to a single atom level, the decomposition of peroxide will be hampered, and it proceeds with peroxide diffusion. Due to this, the reaction proceeds with two electron pathways with the formation of products from the H_2O_2 source. Many noble and non-noble metal SAC catalysts derived from MOFs were explored for ORR reaction. Yu et al. prepared a novel catalyst with single atom-atom grafting of Pt atoms to Fe atomic species in an N-doped carbon matrix [130]. The material exhibited excellent activity in reducing O_2 to H_2O. Because of the stability and reactivity of the substrate and the interaction of Pt-substrate played an important role in good ORR activity. Transition metal-based SAC was also derived based on Fe, Co, Mn, Zn, etc. which showed better ORR performance. Yin and his group encapsulated Co SA in the matrix of N-doped carbon exhibiting an enhanced ORR activity with a half-wave potential of 0.881 V (vs RHE) [131]. Other transition-based metals were also explored for ORR activity. The order of their activities follows the order Fe > Co

> Mn > Cu > Ni in either of the basic and acidic pH mediums. Apart from having SA, additional elements were also doped in many studies to improve the reaction kinetics, stability, and ORR activity due to their unique electronic properties.

5.10.2.2 CO_2 Reduction Reaction

To convert CO_2 into useful products and mitigate the global environmental crisis, SACs were found to be potential candidates due to their high selectivity and activity. The highly active SA sites with unique electronic structures reduces CO_2 to useful value-added products with high conversion efficiency. Mainly, two reactions seemed to be prominent namely CO_2 conversion to CO and CO_2 conversion to alcohols. Ni, Fe, and Co-based SAC were studied in the reduction of CO_2. Wang et al. synthesized Ni SA-loaded carbon matrix with 5.44% of Ni and achieved a faradaic efficiency of 92%–98% over a wide potential range of -0.53 to -1.03 V. The studies confirmed that the active sites for CO reduction were Ni centres and the current density increased with the increase in Ni loading [132]. In a study reported by Wu and their group to understand the effect of Fe-N_4 and Co-N_4 catalysts in CO_2 reduction, Fe sites were proved to be more active than Co sites [133]. To address various issues with individual atomic sites, Zhao et al. encapsulated multiple atomic centres (Fe, Ni) grafted to a porous carbon matrix [134]. The synergistic effect produced by dual atoms gave a conversion efficiency of about 98% and a current density of 74 mA/cm² at a potential of -0.7 V (vs RHE). The durability also remained at 99% up to 30 h of catalytic reaction.

The CO_2 conversion to various C_1 and C_2 products by SAC derived from MOF is still in a fancy stage. However, few studies were reported especially on the Cu SA-based systems where it showed profound activity in the formation of alcoholic products such as methanol. He et al. dispersed Cu SA in the matrix of carbon nanofibers (CNF) for the reduction of CO_2 to methanol with 44% faradaic efficiency. The DFT calculations revealed the pathway for methanol production instead of CH_4 production suggesting the significance of SA. Recent reports suggested that the effect of incorporating multiple SA into the porous substrate, modifying the local coordination structure, and adopting different synthesis routes might help in the CO_2 reduction to various C_1 and C_2 products.

5.10.2.3 Nitrogen Reduction Reaction (NRR)

Nitrogen (N_2) which is abundantly available in the atmosphere can be reduced to NH_3 which is an important material in industries. Industrially NH_3 is synthesized by the Haber-Bosch process which consumes a lot of energy. To avoid this, SAC was used which has high reactivity and conversion efficiency that can bring down the production costs. The electrochemical conversion of N_2 involving several charge transfer sequences and intermediate production seemed to be a viable and facile solution to reduce N_2 to NH_3. NRR follows two different pathways associative and dissociative which involve several reactions as shown in Figure 5.5.

Many factors such as temperature, reaction kinetics, and electrolyte will influence the NRR reactions. Noble metal such as Au was incorporated into the C_3N_4 matrix which showed good activity for NRR reduction [136]. Ru-based SAC were

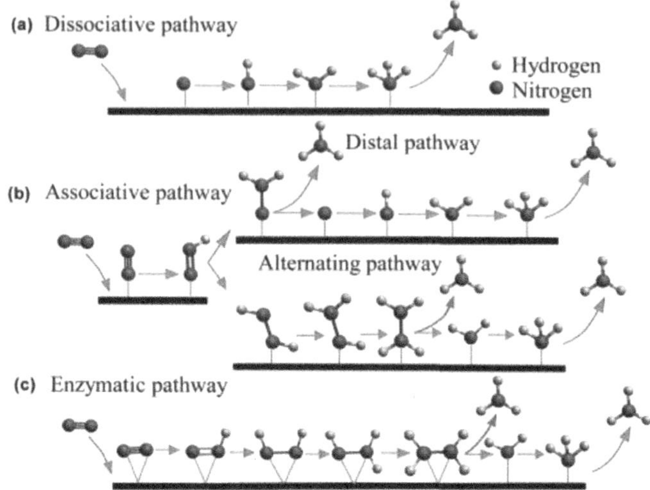

FIGURE 5.5 Possible reduction mechanisms occurring in NRR. Reproduced with permission from Ref. [135].

also prepared by incorporating them into the N-doped carbon structure obtained by pyrolysis which showed 29.6% faradic efficiency and high current density. Generally, side reactions such as HER will take place during the NRR process. To avoid this competition, Tao et al. embedded ZrO_2 into the Ru/N-C structure which showed good NH_3 formation and suppression of HER activity [137]. ZrO_2 contributed to hampering the HER, while Ru sites improved NRR activity. Liu and his group developed a catalyst loaded with Fe SA which showed excellent activity in neutral media. The mechanism follows the transfer of electrons from Fe to N molecule thereby extending the bond length between two N atoms, followed by the N_2 activation procedure. The reaction proceeded in a direction with a low free energy barrier and this neutral media catalytic reaction mimicked the natural N_2 fixation process.

5.10.2.4 Hydrogen Evolution Reaction (HER)

MOF-derived SAC showed promising results in the field of electrocatalytic HER reactions. Transition metal-based SA were incorporated into the carbon framework which enhanced the catalytic activity due to their high activity. Fan and his co-workers synthesized a Ni-C SAC which gave an overpotential of −34 mV and a low Tafel slope of 41 mV/dec at a current density of 10 mA/cm². The Ni SA was strongly coupled to the porous carbon substrate which showed good stability and charge transfer efficiency [138]. Xing and Gao reported an atomically dispersed Ru-doped catalyst derived from ZIF-900 for HER activity which exhibited excellent stability and activity with a low overpotential of 51.6 mV [139]. The novel structure attributed to the Ru site reduced the free energy barrier for the HER reaction. Chen and his group reported a W-SAC encapsulated on an N-doped carbon framework derived from UiO for HER reaction [140]. The strong interaction between the W-SA and the carrier species resulted in excellent activity and it was retained for a longer duration of time. These results paved the way to develop highly active SAC for water-splitting HER reactions.

5.10.2.5 Oxygen Evolution Reaction (OER)

When compared to traditional oxide catalysts such as RuO_2 and IrO_2 which show excellent performance towards OER, SA comprised of these metals show better activity. Combining these SA sites with suitable substrates will incur a huge improvement in performance. For example, Ru SA was encapsulated with a Pt-containing matrix for enhancing the OER activity and it was also observed that the catalyst gained resistance against corrosion [141]. The material exhibited an overpotential of 220 mV at a density of 10 mA/cm^2 and excellent stability was retained even after 28 h of operation. The Pt matrix provided rich defective sites to which Ru SA got anchored, thus giving stability to the catalyst. The unchanged oxidation states of Ru confirmed its corrosion resistance capability. In a recent work, transition metal-based SA sites were coordinated in the N-doped graphitic matrix. The as-prepared catalyst structure which was termed as M-N-C moiety was proven to be best structure by DFT calculations that could achieve remarkable catalytic activity. The metal sites occupied the vacancies present in the graphene structure which was confirmed by various spectroscopic techniques. Three distinct catalysts were prepared with three different metals Co, Ni, and Fe [142]. The OER activity exhibited by them was in the order of Ni > Co > Fe. A novel method of preparation of catalyst where atomic scale CoO_x species generated using O_2 plasma were deposited onto the porous carbon matrix which enhanced the OER activity and outperformed RuO_2 catalysts [143]. The plasma etching broke the Co-N bonds and then the Co reacted with O_2 to form CoO_x.

Likewise, a wide variety of approaches were explored to deposit SA sites in the porous matrix which enable the catalyst to exhibit much higher performance in the field of catalysis. Even though, SAC demonstrated outstanding performance, the ongoing research on SAC is still in developing stage. There is still a research gap in exploring the SAC and efforts must be kept to achieve desired results.

5.11 CONCLUSION

In conclusion, MOF-derived materials seem to be the potential candidates in the field of catalysis since they outweigh their parent MOFs in terms of performance, stability, and cyclability towards catalysis. Several factors pave a way to prefer to MOF-derived materials for their outstanding performance as stated below.

1. The synthesis route is simple and facile, involving an easy pyrolysis route.
2. The inheritance of the structure and morphology of the parent material.
3. Excellent stability and enhanced performance due to the presence of rich active sites.
4. Excellent mass transport and charge transfer due to the underlying high surface area of porous substrate.
5. Prevention of agglomeration of NP by the porous matrix by anchoring them to its defective sites.

6. Able to achieve structural diversity allowing to incorporate many functional guest species during pyrolysis.
7. Able to achieve desired performance by targeting required structure and morphology by varying the synthesis parameters.

As mentioned, MOFs offer wide variety of opportunities to get desired performance in various catalytic applications. For instance, to attain high conductivity for electrocatalytic applications, several conducting species can be incorporated into the MOF precursor. To achieve good mass transport, the structure and morphology can be adjusted by modifying operating parameters. Likewise, multiple range of features can be achieved through MOF-derived materials especially for catalytic applications. Even though, MOF-derived materials offer several advantages, they face challenges that need to be overcome to achieve better results in the field of catalysis. Few of them are the high cost and the incapability of bulk-synthesis, availability of only a few choice of MOF precursors, difficulty in precise control over the dispersion of active sites, etc. Due to these limitations, MOF-derived materials are still in fancy stage and more research must be done to obtain better results.

REFERENCES

1. Deutschmann, O.; Knözinger, H.; Kochloefl, K.; Turek, T. Heterogeneous Catalysis and Solid Catalysts. In *Ullmann's Encyclopedia of Industrial Chemistry*; Wiley-VCH Verlag GmbH & Co. KGaA, 2009. https://doi.org/10.1002/14356007.a05_313.pub2.
2. Ananikov, P.; Nickel, V. The "Spirited Horse" of Transition Metal Catalysis. *ACS Catal*, 2015, *5* (3), 1964–1971. https://doi.org/10.1021/acscatal.5b00072.
3. Robertson, A. J. B. The Early History of Catalysis. *Platinum Met Rev*, 1975, *19*, 64–69.
4. Cordus, V. 1515–1544. *Le guidon des apotiquaires. C'est à dire, la vraye forme et maniere de composer les médicamens... / Traduite de latin en françoys, et repurgée*; Cloquemin, 1575.
5. Augustin, A. Www.e-Rara.Ch Expériences et Réflexions Relatives à l'analyse Du Bled et Des Farines. https://doi.org/10.3931/e-rara-10331.
6. Astruc, D. Introduction: Nanoparticles in Catalysis. *Chem Rev*, 2020, *120* (2), 461–463. https://doi.org/10.1021/acs.chemrev.8b00696.
7. Yang, H. Bin; Miao, J.; Hung, S. F.; Chen, J.; Tao, H. B.; Wang, X.; Zhang, L.; Chen, R.; Gao, J.; Chen, H. M.; et al. Identification of Catalytic Sites for Oxygen Reduction and Oxygen Evolution in N-Doped Graphene Materials: Development of Highly Efficient Metal-Free Bifunctional Electrocatalyst. *Sci Adv*, 2016, *2* (4). https://doi.org/10.1126/sciadv.1501122.
8. Lei, Y.; Wei, L.; Zhai, S.; Wang, Y.; Karahan, H. E.; Chen, X.; Zhou, Z.; Wang, C.; Sui, X.; Chen, Y. Metal-Free Bifunctional Carbon Electrocatalysts Derived from Zeolitic Imidazolate Frameworks for Efficient Water Splitting. *Mater Chem Front*, 2018, *2* (1), 102–111. https://doi.org/10.1039/C7QM00452D.
9. Qian, Y.; Hu, Z.; Ge, X.; Yang, S.; Peng, Y.; Kang, Z.; Liu, Z.; Lee, J. Y.; Zhao, D. A Metal-Free ORR/OER Bifunctional Electrocatalyst Derived from Metal-Organic Frameworks for Rechargeable Zn-Air Batteries. *Carbon N Y*, 2017, *111*, 641–650. https://doi.org/10.1016/j.carbon.2016.10.046.
10. Ai, L.; Tian, T.; Jiang, J. Ultrathin Graphene Layers Encapsulating Nickel Nanoparticles Derived Metal–Organic Frameworks for Highly Efficient Electrocatalytic Hydrogen

and Oxygen Evolution Reactions. *ACS Sustainable Chem Eng*, 2017, *5* (6), 4771–4777. https://doi.org/10.1021/acssuschemeng.7b00153.

11. Wang, C.; Yang, H.; Zhang, Y.; Wang, Q. NiFe Alloy Nanoparticles with Hcp Crystal Structure Stimulate Superior Oxygen Evolution Reaction Electrocatalytic Activity. *Angew Chem Int Ed*, 2019, *58* (18), 6099–6103. https://doi.org/10.1002/anie.201902446.

12. Yi Ma, T.; Dai, S.; Jaroniec, M.; Zhang Qiao, S. Metal–Organic Framework Derived Hybrid Co3O4-Carbon Porous Nanowire Arrays as Reversible Oxygen Evolution Electrodes. *J Am Chem Soc*, 2014, *136* (39), 13925–13931. https://doi.org/10.1021/ja5082553.

13. Yang, X.; Chen, J.; Chen, Y.; Feng, P.; Lai, H.; Li, J.; Luo, X. Novel CO3O4 Nanoparticles/Nitrogen-Doped Carbon Composites with Extraordinary Catalytic Activity for Oxygen Evolution Reaction (OER). *Nanomicro Lett*, 2018, *10* (1), 1–11. https://doi.org/10.1007/s40820-017-0170-4.

14. Hou, Y.; Wen, Z.; Cui, S.; Ci, S.; Mao, S.; Chen, J. An Advanced Nitrogen-Doped Graphene/Cobalt-Embedded Porous Carbon Polyhedron Hybrid for Efficient Catalysis of Oxygen Reduction and Water Splitting. *Adv Funct Mater*, 2015, *25* (6), 872–882. https://doi.org/10.1002/adfm.201403657.

15. Xu, Y.; Tu, W.; Zhang, B.; Yin, S.; Huang, Y.; Kraft, M.; Xu, R. Nickel Nanoparticles Encapsulated in Few-Layer Nitrogen-Doped Graphene Derived from Metal–Organic Frameworks as Efficient Bifunctional Electrocatalysts for Overall Water Splitting. *Adv Mater*, 2017, *29* (11). https://doi.org/10.1002/adma.201605957.

16. Chen, T. T.; Wang, R.; Li, L. K.; Li, Z. J.; Zang, S. Q. MOF-Derived Co9S8/MoS2 Embedded in Tri-Doped Carbon Hybrids for Efficient Electrocatalytic Hydrogen Evolution. *J Energy Chem*, 2020, *44*, 90–96. https://doi.org/10.1016/j.jechem.2019.09.018.

17. Tabassum, H.; Guo, W.; Meng, W.; Mahmood, A.; Zhao, R.; Wang, Q.; Zou, R. Metal–Organic Frameworks Derived Cobalt Phosphide Architecture Encapsulated into B/N Co-Doped Graphene Nanotubes for All PH Value Electrochemical Hydrogen Evolution. *Adv Energy Mater*, 2017, *7* (9). https://doi.org/10.1002/aenm.201601671.

18. Pan, Z.; Pan, N.; Chen, L.; He, J.; Zhang, M. Flower-Like MOF-Derived Co–N-Doped Carbon Composite with Remarkable Activity and Durability for Electrochemical Hydrogen Evolution Reaction. *Int J Hydrogen Energy*, 2019, *44* (57), 30075–30083. https://doi.org/10.1016/j.ijhydene.2019.09.117.

19. Wu, H. Bin; Xia, B. Y.; Yu, L.; Yu, X. Y.; Lou, X. W. Porous Molybdenum Carbide Nano-Octahedrons Synthesized via Confined Carburization in Metal-Organic Frameworks for Efficient Hydrogen Production. *Nat Commun*, 2015, *6*. https://doi.org/10.1038/ncomms7512.

20. Zhao, C.; Dai, X.; Yao, T.; Chen, W.; Wang, X.; Wang, J.; Yang, J.; Wei, S.; Wu, Y.; Li, Y. Ionic Exchange of Metal–Organic Frameworks to Access Single Nickel Sites for Efficient Electroreduction of CO2. *J Am Chem Soc*, 2017, *139* (24), 8078–8081. https://doi.org/10.1021/jacs.7b02736.

21. Cheng, N.; Ren, L.; Xu, X.; Du, Y.; Dou, S. X. Recent Development of Zeolitic Imidazolate Frameworks (ZIFs) Derived Porous Carbon Based Materials as Electrocatalysts. *Adv Energy Mater*. Wiley-VCH Verlag September 5, 2018. https://doi.org/10.1002/aenm.201801257.

22. Albo, J.; Vallejo, D.; Beobide, G.; Castillo, O.; Castaño, P.; Irabien, A. Copper-Based Metal–Organic Porous Materials for CO2 Electrocatalytic Reduction to Alcohols. *ChemSusChem*, 2017, *10* (6), 1100–1109. https://doi.org/10.1002/cssc.201600693.

23. Han, Z.; Kortlever, R.; Chen, H.-Y.; C. Peters, J.; Agapie, T. CO2 Reduction Selective for C≥2 Products on Polycrystalline Copper with N-Substituted Pyridinium Additives. *ACS Cent Sci*, 2017, *3* (8), 853–859. https://doi.org/10.1021/acscentsci.7b00180.

24. Gattrell, M.; Gupta, N.; Co, A. A Review of the Aqueous Electrochemical Reduction of CO2 to Hydrocarbons at Copper. *J Electroanal Chem*. Elsevier August 15, 2006, 1–19. https://doi.org/10.1016/j.jelechem.2006.05.013.

25. Qiao, J.; Liu, Y.; Hong, F.; Zhang, J. A Review of Catalysts for the Electroreduction of Carbon Dioxide to Produce Low-Carbon Fuels. *Chem Soc Rev*. Royal Society of Chemistry January 21, 2014, 631–675. https://doi.org/10.1039/c3cs60323g.

26. Hori, Y.; Murata, A.; Takahashi, R. Formation of Hydrocarbons in the Electrochemical Reduction of Carbon Dioxide at a Copper Electrode in Aqueous Solution. *J Chem Soc, Faraday Trans 1*, 1989, *85* (8), 2309–2326. https://doi.org/10.1039/F1989 8502309.

27. Peterson, A. A.; Abild-Pedersen, F.; Studt, F.; Rossmeisl, J.; Nørskov, J. K. How Copper Catalyzes the Electroreduction of Carbon Dioxide into Hydrocarbon Fuels. *Energy Environ Sci*, 2010, *3* (9), 1311–1315. https://doi.org/10.1039/c0ee00071j.

28. Nam, D.-H.; S. Bushuyev, O.; Li, J.; De Luna, P.; Seifitokaldani, A.; Dinh, C.-T.; Pelayo García de Arquer, F.; Wang, Y.; Liang, Z.; H. Proppe, A.; et al. Metal–Organic Frameworks Mediate Cu Coordination for Selective CO2 Electroreduction. *J Am Chem Soc*, 2018, *140* (36), 11378–11386. https://doi.org/10.1021/jacs.8b06407.

29. Zhao, K.; Liu, Y.; Quan, X.; Chen, S.; Yu, H. CO2 Electroreduction at Low Overpotential on Oxide-Derived Cu/Carbons Fabricated from Metal Organic Framework. *ACS Appl Mater Interfaces*, 2017, *9* (6), 5302–5311. https://doi.org/10.1021/acsami.6b15402.

30. Young Lee, S.; Jung, H.; Kim, N.-K.; Oh, H.-S.; Koun Min, B.; Jeong Hwang, Y. Mixed Copper States in Anodized Cu Electrocatalyst for Stable and Selective Ethylene Production from CO2 Reduction. *J Am Chem Soc*, 2018, *140* (28), 8681–8689. https://doi.org/10.1021/jacs.8b02173.

31. Yao, K.; Xia, Y.; Li, J.; Wang, N.; Han, J.; Gao, C.; Han, M.; Shen, G.; Liu, Y.; Seifitokaldani, A.; et al. Metal-Organic Framework Derived Copper Catalysts for CO2 to Ethylene Conversion. *J Mater Chem A Mater*, 2020, *8* (22), 11117–11123. https://doi.org/10.1039/d0ta02395g.

32. Liu, F.; Wu, C.; Yang, S. Strain and Ligand Effects on CO2 Reduction Reactions over Cu–Metal Heterostructure Catalysts. *J Phys Chem C*, 2017, *121* (40), 22139–22146. https://doi.org/10.1021/acs.jpcc.7b07081.

33. Jansonius, P. R.; Reid, M. L.; Virca, N. C.; Berlinguette, P. C. Strain Engineering Electrocatalysts for Selective CO2 Reduction. *ACS Energy Lett*, 2019, *4* (4), 980–986. https://doi.org/10.1021/acsenergylett.9b00191.

34. Liu, J.; Peng, L.; Zhou, Y.; Lv, L.; Fu, J.; Lin, J.; Guay, D.; Qiao, J. Metal–Organic-Frameworks-Derived Cu/Cu2O Catalyst with Ultrahigh Current Density for Continuous-Flow CO2 Electroreduction. *ACS Sustainable Chem Eng*, 2019, *7* (18), 15739–15746. https://doi.org/10.1021/acssuschemeng.9b03892.

35. Li, D.; Liu, T.; Yan, Z.; Zhen, L.; Liu, J.; Wu, J.; Feng, Y. MOF-Derived Cu2O/Cu Nanospheres Anchored in Nitrogen-Doped Hollow Porous Carbon Framework for Increasing the Selectivity and Activity of Electrochemical CO2-to-Formate Conversion. *ACS Appl Mater Interfaces*, 2020, *12* (6), 7030–7037. https://doi.org/10.1021/acsami.9b15685.

36. Hori, Y.; Wakebe, H.; Tsukamoto, T.; Koga, O. Electrocatalytic Process of CO Selectivity in Electrochemical Reduction of CO2 at Metal Electrodes in Aqueous Media. *Electrochim Acta*, 1994; Vol. 39, 1833–1839.

37. Won, D. H.; Shin, H.; Koh, J.; Chung, J.; Lee, H. S.; Kim, H.; Woo, S. I. Highly Efficient, Selective, and Stable CO2 Electroreduction on a Hexagonal Zn Catalyst. *Angew Chem*, 2016, *128* (32), 9443–9446. https://doi.org/10.1002/ange.201602888.

38. Wang, D.; Zhu, Y.; Yu, W.; He, Z.; Dong, F.; Shen, Y.; Zeng, T.; Lu, X.; Ma, J.; Wang, L.; et al. Ag-MOF-Derived 3D Ag Dendrites Used for the Efficient Electrocatalytic Reduction of CO2 to CO. *Electrochim Acta*, 2022, *403*. https://doi.org/10.1016/j.electa cta.2021.139652.

39. Hu, Y.; Wu, H.; Yang, Y.; Lin, X.; Cheng, H.; Zhang, R.; Jiang, X.; Wang, J. ZIF-8 Derived Porous ZnO with Grain Boundaries for Efficient CO2 Electroreduction. *J Nanopart Res*, 2021, *23* (6). https://doi.org/10.1007/s11051-021-05271-9.

40. Zhao, C.; Dai, X.; Yao, T.; Chen, W.; Wang, X.; Wang, J.; Yang, J.; Wei, S.; Wu, Y.; Li, Y. Ionic Exchange of Metal–Organic Frameworks to Access Single Nickel Sites for Efficient Electroreduction of CO2. *J Am Chem Soc*, 2017, *139* (24), 8078–8081. https://doi.org/10.1021/jacs.7b02736.

41. Wang, X.; Chen, Z.; Zhao, X.; Yao, T.; Chen, W.; You, R.; Zhao, C.; Wu, G.; Wang, J.; Huang, W.; et al. Regulation of Coordination Number over Single Co Sites: Triggering the Efficient Electroreduction of CO2. *Angew Chem Int Ed*, 2018, *57* (7), 1944–1948. https://doi.org/10.1002/anie.201712451.

42. Ye, Y.; Cai, F.; Li, H.; Wu, H.; Wang, G.; Li, Y.; Miao, S.; Xie, S.; Si, R.; Wang, J.; et al. Surface Functionalization of ZIF-8 with Ammonium Ferric Citrate toward High Exposure of Fe-N Active Sites for Efficient Oxygen and Carbon Dioxide Electroreduction. *Nano Energy*, 2017, *38*, 281–289. https://doi.org/10.1016/j.nan oen.2017.05.042.

43. Wang, G.; Sun, Y.; Li, D.; Liang, H.-W.; Dong, R.; Feng, X.; Müllen, K. Controlled Synthesis of N-Doped Carbon Nanospheres with Tailored Mesopores through Self-Assembly of Colloidal Silica. *Angew Chem*, 2015, *127* (50), 15406–15411. https://doi.org/10.1002/ange.201507735.

44. Bechambi, O.; Sayadi, S.; Najjar, W. Photocatalytic Degradation of Bisphenol A in the Presence of C-Doped ZnO: Effect of Operational Parameters and Photodegradation Mechanism. *J Ind Eng Chem*, 2015, *32*, 201–210. https://doi.org/10.1016/j.jiec.2015.08.017.

45. Xia, Q.; Yu, X.; Zhao, H.; Wang, S.; Wang, H.; Guo, Z.; Xing, H. Syntheses of Novel Lanthanide Metal–Organic Frameworks for Highly Efficient Visible-Light-Driven Dye Degradation. *Cryst Growth Des*, 2017, *17* (8), 4189–4195. https://doi.org/10.1021/acs.cgd.7b00504.

46. Yang, T.; Peng, J.; Zheng, Y.; He, X.; Hou, Y.; Wu, L.; Fu, X. Enhanced Photocatalytic Ozonation Degradation of Organic Pollutants by ZnO Modified TiO2 Nanocomposites. *Appl Catal B*, 2018, *221*, 223–234. https://doi.org/10.1016/j.apcatb.2017.09.025.

47. Wang, B.; Wei, K.; Mo, X.; Hu, J.; He, G.; Wang, Y.; Li, W.; He, Q. Improvement in Recycling Times and Photodegradation Efficiency of Core-Shell Structured FeO@C-TiO Composites by PH Adjustment. *ES Materials & Manufacturing*, 2019, *4*, 51–57. https://doi.org/10.30919/esmm5f.

48. Hou, X.; Stanley, S. L.; Zhao, M.; Zhang, J.; Zhou, H.; Cai, Y.; Huang, F.; Wei, Q. MOF-Based C-Doped Coupled TiO2/ZnO Nanofibrous Membrane with Crossed Network Connection for Enhanced Photocatalytic Activity. *J Alloys Compd*, 2019, *777*, 982–990. https://doi.org/10.1016/j.jallcom.2018.10.174.

49. Wang, Y.; Liu, X.; Guo, L.; Shang, L.; Ge, S.; Song, G.; Naik, N.; Shao, Q.; Lin, J.; Guo, Z. Metal Organic Framework-Derived C-Doped ZnO/TiO2 Nanocomposite Catalysts for Enhanced Photodegradation of Rhodamine B. *J Colloid Interface Sci*, 2021, *599*, 566–576. https://doi.org/10.1016/j.jcis.2021.03.167.

50. Wu, R.; Xu, Y.; Xu, R.; Huang, Y.; Zhang, B. Ultrathin-Nanosheet-Based 3D Hierarchical Porous In2S3 Microspheres: Chemical Transformation Synthesis, Characterization, and Enhanced Photocatalytic and Photoelectrochemical Property. *J Mater Chem A Mater*, 2015, *3* (5), 1930–1934. https://doi.org/10.1039/c4ta05729e.

51. An, X.; Yu, J. C.; Wang, F.; Li, C.; Li, Y. One-Pot Synthesis of In2S3 Nanosheets/Graphene Composites with Enhanced Visible-Light Photocatalytic Activity. *Appl Catal B*, 2013, *129*, 80–88. https://doi.org/10.1016/j.apcatb.2012.09.008.

52. Wang, S.; Yuan Guan, B.; Lu, Y.; Wen "David" Lou, X. Formation of Hierarchical In2S3–CdIn2S4 Heterostructured Nanotubes for Efficient and Stable Visible Light CO2 Reduction. *J Am Chem Soc*, 2017, *139* (48), 17305–17308. https://doi.org/10.1021/jacs.7b10733.

53. Wang, H.; Yuan, X.; Wu, Y.; Zeng, G.; Dong, H.; Chen, X.; Leng, L.; Wu, Z.; Peng, L. In Situ Synthesis of In2S3 at MIL-125(Ti) Core-Shell Microparticle for the Removal of Tetracycline from Wastewater by Integrated Adsorption and Visible-Light-Driven Photocatalysis. *Appl Catal B*, 2016, *186*, 19–29. https://doi.org/10.1016/j.apcatb.2015.12.041.

54. Li, G.-C.; Liu, M.; Wu, M.-K.; Liu, P.-F.; Zhou, Z.; Zhu, S.-R.; Liu, R.; Han, L. MOF-Derived Self-Sacrificing Route to Hollow NiS2/ZnS Nanospheres for High Performance Supercapacitors. *RSC Adv*, 2016, *6* (105), 103517–103522. https://doi.org/10.1039/C6RA23071G.

55. Guan, C.; Liu, X.; Elshahawy, A. M.; Zhang, H.; Wu, H.; Pennycook, S. J.; Wang, J. Metal–Organic Framework Derived Hollow CoS2 Nanotube Arrays: An Efficient Bifunctional Electrocatalyst for Overall Water Splitting. *Nanoscale Horiz*, 2017, *2* (6), 342–348. https://doi.org/10.1039/C7NH00079K.

56. Fang, Y.; Zhu, S. R.; Wu, M. K.; Zhao, W. N.; Han, L. MOF-Derived In2S3 Nanorods for Photocatalytic Removal of Dye and Antibiotics. *J Solid State Chem*, 2018, *266*, 205–209. https://doi.org/10.1016/j.jssc.2018.07.026.

57. Qin, H.; Li, W.; Xia, Y.; He, T. Photocatalytic Activity of Heterostructures Based on ZnO and N-Doped ZnO. *ACS Appl Mater Interfaces*, 2011, *3* (8), 3152–3156. https://doi.org/10.1021/am200655h.

58. Shanmugam, V.; Jeyaperumal, K. S. Investigations of Visible Light Driven Sn and Cu Doped ZnO Hybrid Nanoparticles for Photocatalytic Performance and Antibacterial Activity. *Appl Surf Sci*, 2018, *449*, 617–630. https://doi.org/10.1016/j.apsusc.2017.11.167.

59. Lei, X.; Cao, Y.; Chen, Q.; Ao, X.; Fang, Y.; Liu, B. ZIF-8 Derived Hollow CuO/ZnO Material for Study of Enhanced Photocatalytic Performance. *Colloids Surf A Physicochem Eng Asp*, 2019, *568*, 1–10. https://doi.org/10.1016/j.colsurfa.2019.01.072.

60. Shi, X.; Yang, X.; Gu, X.; Su, H. CuO-ZnO Heterometallic Hollow Spheres: Morphology and Defect Structure. *J Solid State Chem*, 2012, *186*, 76–80. https://doi.org/10.1016/j.jssc.2011.11.045.

61. Hassanpour, M.; Safardoust-Hojaghan, H.; Salavati-Niasari, M.; Yeganeh-Faal, A. Nano-Sized CuO/ZnO Hollow Spheres: Synthesis, Characterization and Photocatalytic Performance. *J Mater Sci Mater Electron*, 2017, *28* (19), 14678–14684. https://doi.org/10.1007/s10854-017-7333-4.

62. Jung, S.; Yong, K. Fabrication of CuO–ZnO Nanowires on a Stainless Steel Mesh for Highly Efficient Photocatalytic Applications. *Chem Commun*, 2011, *47* (9), 2643–2645. https://doi.org/10.1039/C0CC04985A.

63. Kumar, D. P.; Reddy, N. L.; Karthik, M.; Neppolian, B.; Madhavan, J.; Shankar, M. V. Solar Light Sensitized P-Ag2O/n-TiO2 Nanotubes Heterojunction Photocatalysts for

Enhanced Hydrogen Production in Aqueous-Glycerol Solution. *Sol Energy Mater Sol Cells*, 2016, *154*, 78–87. https://doi.org/10.1016/j.solmat.2016.04.033.

64. Low, J.; Cao, S.; Yu, J.; Wageh, S. Two-Dimensional Layered Composite Photocatalysts. *Chem Commun*, 2014, *50* (74), 10768–10777. https://doi.org/10.1039/C4CC02553A.

65. Shimidzu, T.; Iyoda, T.; Koide, Y. An Advanced Visible-Light-Induced Water Reduction with Dye-Sensitized Semiconductor Powder Catalyst. *J Am Chem Soc*, 2002, *107* (1), 35–41. https://doi.org/10.1021/ja00287a007.

66. Reddy, D. A.; Kim, H. K.; Kim, Y.; Lee, S.; Choi, J.; Islam, M. J.; Kumar, D. P.; Kim, T. K. Multicomponent Transition Metal Phosphides Derived from Layered Double Hydroxide Double-Shelled Nanocages as an Efficient Non-Precious Co-Catalyst for Hydrogen Production. *J Mater Chem A Mater*, 2016, *4* (36), 13890–13898. https://doi.org/10.1039/C6TA05741A.

67. Bernard Rufus, I.; Ramakrishnan, V.; Viswanathan, B.; Kuriacose, C. J. Rhodium and Rhodium Sulfide Coated Cadmium Sulfide as a Photocatalyst for Photochemical Decomposition of Aqueous Sulfide. *Langmuir*, 2002, *6* (3), 565–567. https://doi.org/10.1021/la00093a008.

68. Gao, M.-R.; Xu, Y.-F.; Jiang, J.; Yu, S.-H. Nanostructured Metal Chalcogenides: Synthesis, Modification, and Applications in Energy Conversion and Storage Devices. *Chem Soc Rev*, 2013, *42* (7), 2986–3017. https://doi.org/10.1039/C2CS35310E.

69. Su, P.; Liu, H.; Jin, Z. Metal Organic Framework-Derived Co3O4/NiCo2O4 Hollow Double-Shell Polyhedrons for Effective Photocatalytic Hydrogen Generation. *Appl Surf Sci*, 2022, *571*. https://doi.org/10.1016/j.apsusc.2021.151288.

70. Zhou, Y.; Zhou, L.; Ni, C.; He, E.; Yu, L.; Li, X. 3D/2D MOF-Derived CoCeOx/g-C3N4 Z-Scheme Heterojunction for Visible Light Photocatalysis: Hydrogen Production and Degradation of Carbamazepine. *J Alloys Compd*, 2022, *890*. https://doi.org/10.1016/j.jallcom.2021.161786.

71. El-Bery, H. M.; Abdelhamid, H. N. Photocatalytic Hydrogen Generation via Water Splitting Using ZIF-67 Derived Co3O4@C/TiO2. *J Environ Chem Eng*, 2021, *9* (4). https://doi.org/10.1016/j.jece.2021.105702.

72. Wang, X. li; Xiao, Y.; Yu, H.; Yang, Y.; Dong, X. ting; Xia, L. Noble-Metal-Free MOF Derived ZnS/CeO2 Decorated with CuS Cocatalyst Photocatalyst with Efficient Photocatalytic Hydrogen Production Character. *ChemCatChem*, 2020, *12* (22), 5669–5678. https://doi.org/10.1002/cctc.202001118.

73. Ola, O.; Maroto-Valer, M. M. Review of Material Design and Reactor Engineering on TiO2 Photocatalysis for CO2 Reduction. *J Photochem Photobiol, C.* Elsevier B.V. September 1, 2015, 16–42. https://doi.org/10.1016/j.jphotochemrev.2015.06.001.

74. Inoue, T.; Fujishima, A.; Konishi, S.; Honda, K. Photoelectrocatalytic Reduction of Carbon Dioxide in Aqueous Suspensions of Semiconductor Powders. *Nature*, 1979, *277* (5698), 637–638. https://doi.org/10.1038/277637a0.

75. Li, Q.; Gao, Y.; Zhang, M.; Gao, H.; Chen, J.; Jia, H. Efficient Infrared-Light-Driven Photothermal CO2 Reduction over MOF-Derived Defective Ni/TiO2. *Appl Catal B*, 2022, *303*. https://doi.org/10.1016/j.apcatb.2021.120905.

76. Lippi, R.; Howard, S. C.; Barron, H.; Easton, C. D.; Madsen, I. C.; Waddington, L. J.; Vogt, C.; Hill, M. R.; Sumby, C. J.; Doonan, C. J.; et al. Highly Active Catalyst for CO2 Methanation Derived from a Metal Organic Framework Template. *J Mater Chem A Mater*, 2017, *5* (25), 12990–12997. https://doi.org/10.1039/C7TA00958E.

77. Wei, L.; Shifu, C. Preparation and Characterization of P-n Heterojunction Photocatalyst Cu2O/In2O3 and Its Photocatalytic Activity under Visible and UV Light Irradiation. *J Electrochem Soc*, 2010, *157* (11), H1029. https://doi.org/10.1149/1.3489945.

78. Zhang, H.; Wang, T.; Wang, J.; Liu, H.; Dao, T. D.; Li, M.; Liu, G.; Meng, X.; Chang, K.; Shi, L.; et al. Surface-Plasmon-Enhanced Photodriven CO2Reduction Catalyzed by Metal-Organic-Framework-Derived Iron Nanoparticles Encapsulated by Ultrathin Carbon Layers. *Adv Mater*, 2016, *28* (19), 3703–3710. https://doi.org/10.1002/adma.201505187.

79. Khaletskaya, K.; Pougin, A.; Medishetty, R.; Rösler, C.; Wiktor, C.; Strunk, J.; A. Fischer, R. Fabrication of Gold/Titania Photocatalyst for CO2 Reduction Based on Pyrolytic Conversion of the Metal–Organic Framework NH2-MIL-125(Ti) Loaded with Gold Nanoparticles. *Chem Mater*, 2015, *27* (21), 7248–7257. https://doi.org/10.1021/acs.chemmater.5b03017.

80. Liu, Q.; Low, Z.-X.; Li, L.; Razmjou, A.; Wang, K.; Yao, J.; Wang, H. ZIF-8/Zn2GeO4 Nanorods with an Enhanced CO2 Adsorption Property in an Aqueous Medium for Photocatalytic Synthesis of Liquid Fuel. *J Mater Chem A Mater*, 2013, *1* (38), 11563–11569. https://doi.org/10.1039/C3TA12433A.

81. Wang, S.; Wang, Y.; Zang, S. Q.; Lou, X. W. Hierarchical Hollow Heterostructures for Photocatalytic CO2 Reduction and Water Splitting. *Small Methods*. John Wiley and Sons Inc. January 1, 2020. https://doi.org/10.1002/smtd.201900586.

82. Wang, Y.; Wang, S.; Zhang, S. L.; Lou, X. W. Formation of Hierarchical FeCoS2–CoS2 Double-Shelled Nanotubes with Enhanced Performance for Photocatalytic Reduction of CO2. *Angew Chem Int Ed*, 2020, *59* (29), 11918–11922. https://doi.org/10.1002/anie.202004609.

83. Li, K.; Peng, B.; Peng, T. Recent Advances in Heterogeneous Photocatalytic CO2 Conversion to Solar Fuels. *ACS Catal*, 2016, *6* (11), 7485–7527. https://doi.org/10.1021/acscatal.6b02089.

84. Han, C.; Zhang, X.; Huang, S.; Hu, Y.; Yang, Z.; Li, T. T.; Li, Q.; Qian, J. MOF-on-MOF-Derived Hollow Co3O4/In2O3 Nanostructure for Efficient Photocatalytic CO2 Reduction. *Adv Sci*, 2023. https://doi.org/10.1002/advs.202300797.

85. Tan, J.; Yu, M.; Cai, Z.; Lou, X.; Wang, J.; Li, Z. MOF-Derived Synthesis of MnS/In2S3 p-n Heterojunctions with Hierarchical Structures for Efficient Photocatalytic CO2 Reduction. *J Colloid Interface Sci*, 2021, *588*, 547–556. https://doi.org/10.1016/j.jcis.2020.12.110.

86. Yao, X.; Bai, C.; Chen, J.; Li, Y. Efficient and Selective Green Oxidation of Alcohols by MOF-Derived Magnetic Nanoparticles as a Recoverable Catalyst. *RSC Adv*, 2016, *6* (32), 26921–26928. https://doi.org/10.1039/c6ra01617k.

87. Yu, G.; Sun, J.; Muhammad, F.; Wang, P.; Zhu, G. Cobalt-Based Metal Organic Framework as Precursor to Achieve Superior Catalytic Activity for Aerobic Epoxidation of Styrene. *RSC Adv*, 2014, *4* (73), 38804–38811. https://doi.org/10.1039/c4ra03746d.

88. Lin, X.; Liu, B.; Huang, H.; Shi, C.; Liu, Y.; Kang, Z. One-Step Synthesis of ZnS-N/C Nanocomposites Derived from Zn-Based Chiral Metal-Organic Frameworks with Highly Efficient Photocatalytic Activity for the Selective Oxidation of: Cis-Cyclooctene. *Inorg Chem Front*, 2018, *5* (3), 723–731. https://doi.org/10.1039/c7qi00693d.

89. Wang, X.; Li, Y. Nanoporous Carbons Derived from MOFs as Metal-Free Catalysts for Selective Aerobic Oxidations. *J Mater Chem A Mater*, 2016, *4* (14), 5247–5257. https://doi.org/10.1039/c6ta00324a.

90. Kim, B. R.; Oh, J. S.; Kim, J.; Lee, C. Y. Robust Aerobic Alcohol Oxidation Catalyst Derived from Metal-Organic Frameworks. *Catal Letters*, 2016, *146* (4), 734–743. https://doi.org/10.1007/s10562-016-1700-2.

91. Zhong, W.; Liu, H.; Bai, C.; Liao, S.; Li, Y. Base-Free Oxidation of Alcohols to Esters at Room Temperature and Atmospheric Conditions Using Nanoscale Co-Based Catalysts. *ACS Catal*, 2015, *5* (3), 1850–1856. https://doi.org/10.1021/cs502101c.

92. Zhang, S.; Han, A.; Zhai, Y.; Zhang, J.; Cheong, W. C.; Wang, D.; Li, Y. ZIF-Derived Porous Carbon Supported Pd Nanoparticles within Mesoporous Silica Shells: Sintering- and Leaching-Resistant Core-Shell Nanocatalysts. *Chem Commun*, 2017, *53* (68), 9490–9493. https://doi.org/10.1039/c7cc04926a.

93. Zhou, Y. X.; Chen, Y. Z.; Cao, L.; Lu, J.; Jiang, H. L. Conversion of a Metal-Organic Framework to N-Doped Porous Carbon Incorporating Co and CoO Nanoparticles: Direct Oxidation of Alcohols to Esters. *Chem Commun*, 2015, *51* (39), 8292–8295. https://doi.org/10.1039/c5cc01588j.

94. Hui, J.; Chu, H.; Zhang, W.; Shen, Y.; Chen, W.; Hu, Y.; Liu, W.; Gao, C.; Guo, S.; Xiao, G.; et al. Multicomponent Metal-Organic Framework Derivatives for Optimizing the Selective Catalytic Performance of Styrene Epoxidation Reaction. *Nanoscale*, 2018, *10* (18), 8772–8778. https://doi.org/10.1039/c8nr01336e.

95. Wang, F.; He, X.; Sun, L.; Chen, J.; Wang, X.; Xu, J.; Han, X. Engineering an N-Doped TiO2@N-Doped C Butterfly-Like Nanostructure with Long-Lived Photo-Generated Carriers for Efficient Photocatalytic Selective Amine Oxidation. *J Mater Chem A Mater*, 2018, *6* (5), 2091–2099. https://doi.org/10.1039/c7ta09166d.

96. Bai, C.; Yao, X.; Li, Y. Easy Access to Amides through Aldehydic C–H Bond Functionalization Catalyzed by Heterogeneous Co-Based Catalysts. *ACS Catal*, 2015, *5* (2), 884–891. https://doi.org/10.1021/cs501822r.

97. Xu, Q.; Feng, B.; Ye, C.; Fu, Y.; Chen, D. L.; Zhang, F.; Zhang, J.; Zhu, W. Atomically Dispersed Vanadium Sites Anchored on N-Doped Porous Carbon for the Efficient Oxidative Coupling of Amines to Imines. *ACS Appl Mater Interfaces*, 2021, *13* (13), 15168–15177. https://doi.org/10.1021/acsami.0c22453.

98. Qiu, X.; Wang, X.; Li, Y. Controlled Growth of Dense and Ordered Metal-Organic Framework Nanoparticles on Graphene Oxide. *Chem Commun*, 2015, *51* (18), 3874–3877. https://doi.org/10.1039/c4cc09933h.

99. Zhou, Y.; Long, J.; Li, Y. Ni-Based Catalysts Derived from a Metal-Organic Framework for Selective Oxidation of Alkanes. *Cuihua Xuebao/Chin J Catal*, 2016, *37* (6), 955–962. https://doi.org/10.1016/S1872-2067(15)61067-1.

100. Long, J.; Zhou, Y.; Li, Y. Transfer Hydrogenation of Unsaturated Bonds in the Absence of Base Additives Catalyzed by a Cobalt-Based Heterogeneous Catalyst. *Chem Commun*, 2015, *51* (12), 2331–2334. https://doi.org/10.1039/c4cc08946d.

101. Long, J.; Shen, K.; Chen, L.; Li, Y. Multimetal-MOF-Derived Transition Metal Alloy NPs Embedded in an N-Doped Carbon Matrix: Highly Active Catalysts for Hydrogenation Reactions. *J Mater Chem A Mater*, 2016, *4* (26), 10254–10262. https://doi.org/10.1039/c6ta00157b.

102. Li, A.; Shen, K.; Chen, J.; Li, Z.; Li, Y. Highly Selective Hydrogenation of Phenol to Cyclohexanol over MOF-Derived Non-Noble Co-Ni@NC Catalysts. *Chem Eng Sci*, 2017, *166*, 66–76. https://doi.org/10.1016/j.ces.2017.03.027.

103. Jagadeesh, R. V.; Murugesan, K.; Alshammari, A. S.; Neumann, H.; Pohl, M. M.; Radnik, J.; Beller, M. MOF-Derived Cobalt Nanoparticles Catalyze a General Synthesis of Amines. *Science (1979)*, 2017, *358* (6361), 326–332. https://doi.org/10.1126/science.aan6245.

104. Murugesan, K.; Senthamarai, T.; Sohail, M.; Alshammari, A. S.; Pohl, M. M.; Beller, M.; Jagadeesh, R. V. Cobalt-Based Nanoparticles Prepared from MOF-Carbon Templates as Efficient Hydrogenation Catalysts. *Chem Sci*, 2018, *9* (45), 8553–8560. https://doi.org/10.1039/c8sc02807a.

105. Li, Y.; Zhou, Y. X.; Ma, X.; Jiang, H. L. A Metal-Organic Framework-Templated Synthesis of γ-Fe2O3 Nanoparticles Encapsulated in Porous Carbon for Efficient and Chemoselective Hydrogenation of Nitro Compounds. *Chem Commun*, 2016, *52* (22), 4199–4202. https://doi.org/10.1039/c6cc00011h.

106. Ma, X.; Zhou, Y. X.; Liu, H.; Li, Y.; Jiang, H. L. A MOF-Derived Co-CoO@N-Doped Porous Carbon for Efficient Tandem Catalysis: Dehydrogenation of Ammonia Borane and Hydrogenation of Nitro Compounds. *Chem Commun*, 2016, *52* (49), 7719–7722. https://doi.org/10.1039/c6cc03149h.

107. Van Nguyen, C.; Lee, S.; Chung, Y. G.; Chiang, W. H.; Wu, K. C. W. Synergistic Effect of Metal-Organic Framework-Derived Boron and Nitrogen Heteroatom-Doped Three-Dimensional Porous Carbons for Precious-Metal-Free Catalytic Reduction of Nitroarenes. *Appl Catal B*, 2019, *257*. https://doi.org/10.1016/j.apc atb.2019.117888.

108. Liu, W.; Li, S. Q.; Liu, W. X.; Zhang, Q.; Shao, J.; Tian, J. L. MOF-Derived B, N Co-Doped Porous Carbons as Metal-Free Catalysts for Highly Efficient Nitro Aromatics Reduction. *J Environ Chem Eng*, 2021, *9* (4). https://doi.org/10.1016/j.jece.2021.105689.

109. Wang, X.; Zhao, S.; Zhang, Y.; Wang, Z.; Feng, J.; Song, S.; Zhang, H. CeO2 Nanowires Self-Inserted into Porous Co3O4 Frameworks as High-Performance "Noble Metal Free" Hetero-Catalysts. *Chem Sci*, 2016, *7* (2), 1109–1114. https://doi.org/10.1039/c5sc03430b.

110. Li, X.; Zhang, W.; Liu, Y.; Li, R. Palladium Nanoparticles Immobilized on Magnetic Porous Carbon Derived from ZIF-67 as Efficient Catalysts for the Semihydrogenation of Phenylacetylene under Extremely Mild Conditions. *ChemCatChem*, 2016, *8* (6), 1111–1118. https://doi.org/10.1002/cctc.201501283.

111. Nakatsuka, K.; Yoshii, T.; Kuwahara, Y.; Mori, K.; Yamashita, H. Controlled Pyrolysis of Ni-MOF-74 as a Promising Precursor for the Creation of Highly Active Ni Nanocatalysts in Size-Selective Hydrogenation. *Chem Eur J*, 2018, *24* (4), 898–905. https://doi.org/10.1002/chem.201704341.

112. Huang, G.; Yang, L.; Ma, X.; Jiang, J.; Yu, S. H.; Jiang, H. L. Metal-Organic Framework-Templated Porous Carbon for Highly Efficient Catalysis: The Critical Role of Pyrrolic Nitrogen Species. *Chem Eur J*, 2016, *22* (10), 3470–3477. https://doi.org/10.1002/chem.201504867.

113. Shen, K.; Chen, L.; Long, J.; Zhong, W.; Li, Y. MOFs-Templated Co@Pd Core–Shell NPs Embedded in N-Doped Carbon Matrix with Superior Hydrogenation Activities. *ACS Catal*, 2015, *5* (9), 5264–5271. https://doi.org/10.1021/acscatal.5b00998.

114. Yang, H.; Bradley, J. S.; Chan, A.; Waterhouse, I. N. G.; Nann, T.; Kruger, E. P.; Telfer, G. S. Catalytically Active Bimetallic Nanoparticles Supported on Porous Carbon Capsules Derived from Metal–Organic Framework Composites. *J Am Chem Soc*, 2016, *138* (36), 11872–11881. https://doi.org/10.1021/jacs.6b06736.

115. Gu, Z.; Chen, L.; Li, X.; Chen, L.; Zhang, Y.; Duan, C. NH2-MIL-125(Ti)-Derived Porous Cages of Titanium Oxides to Support Pt-Co Alloys for Chemoselective Hydrogenation Reactions. *Chem Sci*, 2019, *10* (7), 2111–2117. https://doi.org/10.1039/c8sc05450a.

116. Ding, M.; Chen, S.; Liu, X. Q.; Sun, L. B.; Lu, J.; Jiang, H. L. Metal–Organic Framework-Templated Catalyst: Synergy in Multiple Sites for Catalytic CO2 Fixation. *ChemSusChem*, 2017, *10* (9), 1898–1903. https://doi.org/10.1002/cssc.201700245.

117. Toyao, T.; Fujiwaki, M.; Miyahara, K.; Kim, T. H.; Horiuchi, Y.; Matsuoka, M. Design of Zeolitic Imidazolate Framework Derived Nitrogen-Doped Nanoporous Carbons

Containing Metal Species for Carbon Dioxide Fixation Reactions. *ChemSusChem*, 2015, *8* (22), 3905–3912. https://doi.org/10.1002/cssc.201500780.

118. Zhang, S.; Liu, H.; Sun, C.; Liu, P.; Li, L.; Yang, Z.; Feng, X.; Huo, F.; Lu, X. CuO/Cu2O Porous Composites: Shape and Composition Controllable Fabrication Inherited from Metal Organic Frameworks and Further Application in CO Oxidation. *J Mater Chem A Mater*, 2015, *3* (10), 5294–5298. https://doi.org/10.1039/c5t a00249d.

119. Li, S.; Wang, N.; Yue, Y.; Wang, G.; Zu, Z.; Zhang, Y. Copper Doped Ceria Porous Nanostructures towards a Highly Efficient Bifunctional Catalyst for Carbon Monoxide and Nitric Oxide Elimination. *Chem Sci*, 2015, *6* (4), 2495–2500. https://doi.org/ 10.1039/c5sc00129c.

120. Liu, H.; Zhang, S.; Liu, Y.; Yang, Z.; Feng, X.; Lu, X.; Huo, F. Well-Dispersed and Size-Controlled Supported Metal Oxide Nanoparticles Derived from MOF Composites and Further Application in Catalysis. *Small*, 2015, *11* (26), 3130–3134. https://doi.org/ 10.1002/smll.201401791.

121. Teng, M.; Luo, L.; Yang, X. Synthesis of Mesoporous Ce1-XZrxO2 (x = 0.2-0.5) and Catalytic Properties of CuO Based Catalysts. *Microporous Mesoporous Mater*, 2009, *119* (1–3), 158–164. https://doi.org/10.1016/j.micromeso.2008.10.019.

122. Zhu, C.; Ding, T.; Gao, W.; Ma, K.; Tian, Y.; Li, X. CuO/CeO2 Catalysts Synthesized from Ce-UiO-66 Metal-Organic Framework for Preferential CO Oxidation. *Int J Hydrogen Energy*, 2017, *42* (27), 17457–17465. https://doi.org/10.1016/j.ijhyd ene.2017.02.088.

123. Wang, X.; Zhong, W.; Li, Y. Nanoscale Co-Based Catalysts for Low-Temperature CO Oxidation. *Catal Sci Technol*, 2015, *5* (2), 1014–1020. https://doi.org/10.1039/c4c y01147c.

124. Santos, V. P.; Wezendonk, T. A.; Jaén, J. J. D.; Dugulan, A. I.; Nasalevich, M. A.; Islam, H. U.; Chojecki, A.; Sartipi, S.; Sun, X.; Hakeem, A. A.; et al. Metal Organic Framework-Mediated Synthesis of Highly Active and Stable Fischer-Tropsch Catalysts. *Nat Commun*, 2015, *6*. https://doi.org/10.1038/ncomms7451.

125. Qiu, B.; Yang, C.; Guo, W.; Xu, Y.; Liang, Z.; Ma, D.; Zou, R. Highly Dispersed Co-Based Fischer-Tropsch Synthesis Catalysts from Metal-Organic Frameworks. *J Mater Chem A Mater*, 2017, *5* (17), 8081–8086. https://doi.org/10.1039/c7t a02128c.

126. Ning, F.; Hou, H.; Morse, A. N.; Lash, G. E. Understanding and Management of Gestational Trophoblastic Disease [Version 1; Peer Review: 2 Approved]. *F1000Research*. F1000 Research Ltd 2019. https://doi.org/10.12688/f1000resea rch.14953.1.

127. Zhang, J.; An, B.; Hong, Y.; Meng, Y.; Hu, X.; Wang, C.; Lin, J.; Lin, W.; Wang, Y. Pyrolysis of Metal-Organic Frameworks to Hierarchical Porous Cu/Zn-Nanoparticle@ carbon Materials for Efficient CO2 Hydrogenation. *Mater Chem Front*, 2017, *1* (11), 2405–2409. https://doi.org/10.1039/c7qm00328e.

128. Gandara Loe, J.; Pinzón Peña, A.; Martin Espejo, J. L.; Bobadilla, L. F.; Ramírez Reina, T.; Pastor-Pérez, L. MIL-100(Fe)-Derived Catalysts for CO2 Conversion via Low- and High-Temperature Reverse Water-Gas Shift Reaction. *Heliyon*, 2023, *9* (5). https://doi.org/10.1016/j.heliyon.2023.e16070.

129. Ronda-Lloret, M.; Rico-Francés, S.; Sepúlveda-Escribano, A.; Ramos-Fernandez, E. V. CuOx/CeO2 Catalyst Derived from Metal Organic Framework for Reverse Water-Gas Shift Reaction. *Appl Catal A Gen*, 2018, *562*, 28–36. https://doi.org/10.1016/j.apcata.2018.05.024.

130. Zeng, X.; Shui, J.; Liu, X.; Liu, Q.; Li, Y.; Shang, J.; Zheng, L.; Yu, R. Single-Atom to Single-Atom Grafting of Pt1 onto Fe—N4 Center : Pt1@Fe—N—C Multifunctional Electrocatalyst with Significantly Enhanced Properties. *Adv Energy Mater*, 2018, *8* (1). https://doi.org/10.1002/aenm.201701345.

131. Yin, P.; Yao, T.; Wu, Y.; Zheng, L.; Lin, Y.; Liu, W.; Ju, H.; Zhu, J.; Hong, X.; Deng, Z.; et al. Single Cobalt Atoms with Precise N-Coordination as Superior Oxygen Reduction Reaction Catalysts. *Angew Chem*, 2016, *128* (36), 10958–10963. https://doi.org/10.1002/ange.201604802.

132. Yan, C.; Li, H.; Ye, Y.; Wu, H.; Cai, F.; Si, R.; Xiao, J.; Miao, S.; Xie, S.; Yang, F.; et al. Coordinatively Unsaturated Nickel-Nitrogen Sites towards Selective and High-Rate CO_2 Electroreduction. *Energy Environ Sci*, 2018, *11* (5), 1204–1210. https://doi.org/10.1039/c8ee00133b.

133. Huang, P.; Huang, J.; Pantovich, A. S.; Carl, D. A.; Fenton, G. T.; Caputo, A. C.; Grimm, L. R.; Frenkel, I. A.; Li, G. Selective CO_2 Reduction Catalyzed by Single Cobalt Sites on Carbon Nitride under Visible-Light Irradiation. *J Am Chem Soc*, 2018, *140* (47), 16042–16047. https://doi.org/10.1021/jacs.8b10380.

134. Ren, W.; Tan, X.; Yang, W.; Jia, C.; Xu, S.; Wang, K.; Smith, S. C.; Zhao, C. Isolated Diatomic Ni-Fe Metal–Nitrogen Sites for Synergistic Electroreduction of CO_2. *Angew Chem Int Ed*, 2019, *58* (21), 6972–6976. https://doi.org/10.1002/anie.201901575.

135. Rehman, F.; Delowar Hossain, M.; Tyagi, A.; Lu, D.; Yuan, B.; Luo, Z. Engineering Electrocatalyst for Low-Temperature N_2 Reduction to Ammonia. *Mater Today*, 2021, *44*, 136–167. https://doi.org/10.1016/j.mattod.2020.09.006.

136. Wang, X.; Wang, W.; Qiao, M.; Wu, G.; Chen, W.; Yuan, T.; Xu, Q.; Chen, M.; Zhang, Y.; Wang, X.; et al. Atomically Dispersed Au1 Catalyst towards Efficient Electrochemical Synthesis of Ammonia. *Sci Bull (Beijing)*, 2018, *63* (19), 1246–1253. https://doi.org/10.1016/j.scib.2018.07.005.

137. Tao, H.; Choi, C.; Ding, L. X.; Jiang, Z.; Han, Z.; Jia, M.; Fan, Q.; Gao, Y.; Wang, H.; Robertson, A. W.; et al. Nitrogen Fixation by Ru Single-Atom Electrocatalytic Reduction. *Chem*, 2019, *5* (1), 204–214. https://doi.org/10.1016/j.chempr.2018.10.007.

138. Fan, L.; Liu, P. F.; Yan, X.; Gu, L.; Yang, Z. Z.; Yang, H. G.; Qiu, S.; Yao, X. Atomically Isolated Nickel Species Anchored on Graphitized Carbon for Efficient Hydrogen Evolution Electrocatalysis. *Nat Commun*, 2016, *7*. https://doi.org/10.1038/ncomms10667.

139. Xing, L.; Gao, H.; Hai, G.; Tao, Z.; Zhao, J.; Jia, D.; Chen, X.; Han, M.; Hong, S.; Zheng, L.; et al. Atomically Dispersed Ruthenium Sites on Whisker-Like Secondary Microstructure of Porous Carbon Host toward Highly Efficient Hydrogen Evolution. *J Mater Chem A Mater*, 2020, *8* (6), 3203–3210. https://doi.org/10.1039/c9ta11280d.

140. Chen, W.; Pei, J.; He, C. T.; Wan, J.; Ren, H.; Wang, Y.; Dong, J.; Wu, K.; Cheong, W. C.; Mao, J.; et al. Single Tungsten Atoms Supported on MOF-Derived N-Doped Carbon for Robust Electrochemical Hydrogen Evolution. *Adv Mater*, 2018, *30* (30). https://doi.org/10.1002/adma.201800396.

141. Yao, Y.; Hu, S.; Chen, W.; Huang, Z. Q.; Wei, W.; Yao, T.; Liu, R.; Zang, K.; Wang, X.; Wu, G.; et al. Engineering the Electronic Structure of Single Atom Ru Sites via Compressive Strain Boosts Acidic Water Oxidation Electrocatalysis. *Nat Catal*, 2019, *2* (4), 304–313. https://doi.org/10.1038/s41929-019-0246-2.

142. Fei, H.; Dong, J.; Feng, Y.; Allen, C. S.; Wan, C.; Volosskiy, B.; Li, M.; Zhao, Z.; Wang, Y.; Sun, H.; et al. General Synthesis and Definitive Structural Identification of MN4C4 Single-Atom Catalysts with Tunable Electrocatalytic Activities. *Nat Catal*, 2018, *1* (1), 63–72. https://doi.org/10.1038/s41929-017-0008-y.

143. Dou, S.; Dong, C. L.; Hu, Z.; Huang, Y. C.; Chen, J. L.; Tao, L.; Yan, D.; Chen, D.; Shen, S.; Chou, S.; et al. Atomic-Scale CoOx Species in Metal–Organic Frameworks for Oxygen Evolution Reaction. *Adv Funct Mater*, 2017, *27* (36). https://doi.org/10.1002/adfm.201702546.

6 Sensing

6.1 INTRODUCTION

In analytical science, the continuous real-time monitoring of various analytes specifically having environmental significance is of great research interest. The development of miniaturized measuring devices termed 'sensors' designed to interact specifically with an analyte connect the different domains of materials science and technology. Basically, a sensor is a device or system that detect the changes in the environment such as light, heat, pressure, moisture, and motion to produce a quantifiable signal which is electronically transmitted to a measuring or control instrument. In other words, a sensor converts a physical phenomenon into a digital signal which can be measured, read and processed. The classification of sensors can be done in a number of ways: based on detection methods, conversion of input-output signal, external excitation signal, etc. Depending on the detection methods sensors can be classified into chemical, biological, electric, radioactive, etc., whereas the signal conversion technique divide the class of sensors into electrochemical, photoelectric, thermoelectric, electromagnetic, thermo-optic and so on. Another broad category is the class of active and passive sensors. Active sensors are those that require an external power source to operate which includes GPS and radar. On the other hand, passive sensors do not require an external power source and they generate their own electric signal to operate. Examples are electric field, thermal, infrared, photographic, seismic sensors and so on. The discussion of all types of sensors are beyond the scope of this book and hence the focus will be on chemical sensors reported in the material science research.

In a general sense, the different parts of a sensor includes a sensing unit, processing unit, transducer and power source. In the analytical point of view, the important characteristics of a sensor are sensitivity, selectivity, stability and low cost, which is determined by the selection and rational design of sensing materials. With the development of material science, a wide range of materials have been used for the past several decades to develop sensors with high performance towards various analytes of interest. Some of the examples are, carbon-based materials (graphene, carbon nanotube, reduced graphene oxide), quantum dots, metal/metal oxide nanoparticles, magnetic nanomaterials, metal–organic frameworks, etc. Among these

DOI: 10.1201/9781003432357-8

broad choice of materials, MOFs have a significant place in the development of novel sensing platforms due to their properties like high surface area, porosity, presence of coordinatively unsaturated sites, ease of functionalization, etc. However, the use of pristine MOF materials for sensor development faces certain challenges due to their inferior electrical conductivity and low chemical stability in aqueous medium. The emerging class of MOF-derived materials can address these challenges and researchers are actively exploring its potential in the development of different types of sensors.

6.2   SENSOR – DEFINITION

Generally, a sensor consists of a chemically selective layer called recognition element/receptor and a signal transducer. The former undergoes a physical change upon interaction with the analyte which is converted to an electric signal by the transducer. Specifically, a chemical sensor detect the presence of a component or its concentration and transform this chemical information into an analytical signal (Figure 6.1).

Compared to other established sophisticated analytical apparatus such as spectrometry, chromatographic techniques, etc. sensors possess advantages like higher sensitivity, selectivity, ease of automation, portability, low cost, miniaturization, stability, minimum sample amount requirement, etc.

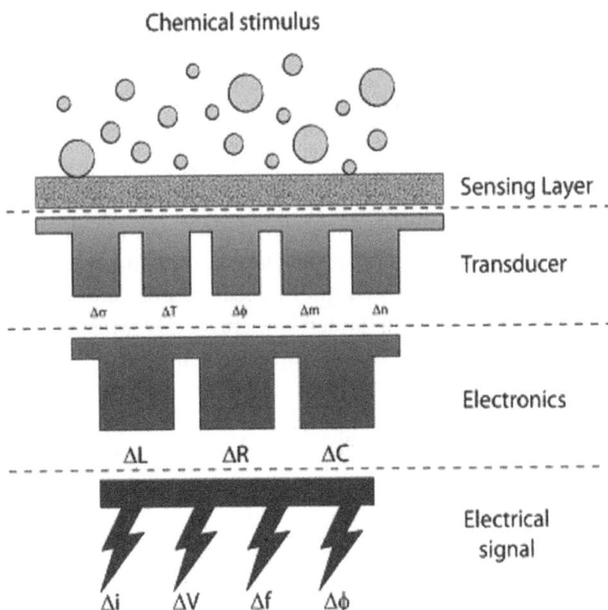

FIGURE 6.1   Schematic representation of a sensor. Reproduced with permission from Ref. [1].

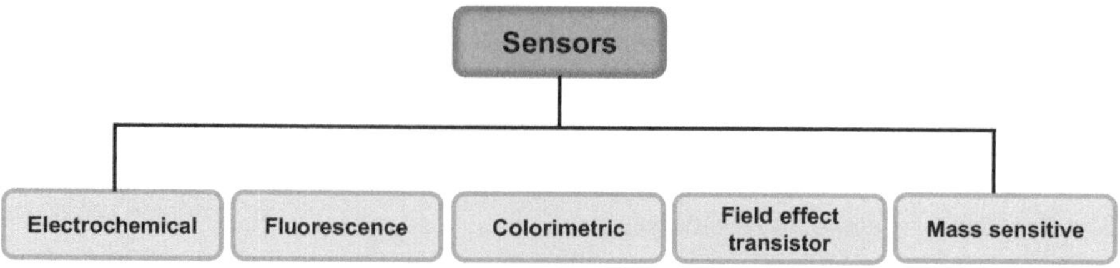

FIGURE 6.2   Schematic diagram of the classification of chemical sensors.

6.3   CLASSIFICATION OF SENSOR

In a general sense, sensors can be broadly classified into five different categories based on the underlying principle of sensing mechanism (Figure 6.2).

6.4   ANALYTES OF INTEREST

The analytes of interest in sensing studies can vary widely depending on the specific application and the information being sought. These analytes may encompass a broad range of substances, from chemical compounds to biological molecules (Table 6.1).

These are some examples of the many analytes of interest in sensing studies. Researchers develop specific sensors and sensing techniques to detect and quantify these analytes in diverse applications ranging from healthcare and environmental monitoring to industrial quality control and scientific research. The choice of analyte and sensor technology depends on the goals of the study and the desired sensitivity and specificity.

6.5   ELECTROCHEMICAL SENSOR

Electrochemical sensors are devices which provide the composition of a system in real time by coupling a chemically selective layer/recognition element to an electrochemical transducer. If the recognition element is a biomolecule, such as enzyme, aptamer or antibodies immobilized on the electrode surface, then the system is called an electrochemical biosensor. Generally, electrochemical sensors can be broadly classified into four categories based on the underlying principle of sensing: voltammetric, potentiometric, conductometric and amperometric sensors.

- In voltammetric sensors, the applied potential is varied continuously or in a step-wise manner and the resultant current is measured as a function of potential. The different voltammetric techniques used to study the electrochemistry of active species in solution include: cyclic voltammetry (CV), linear sweep voltammetry (LSV), and pulse voltammetry. Pulsed voltammetry consist of differential pulse voltammetry (DPV) and square wave voltammetry (SWV).
- Potentiometric sensors measure the potential difference between two electrodes when there is no current flow. The obtained potential is used to determine the concentration of components/analytes in solution. Ion selective electrode (ISE)

TABLE 6.1
Classification of analytes used for sensing

Analyte classification	Category	Examples
Chemical	Gases	VOCs, CO_2, O_2, N_2, NH_3
	Ions	Metal/heavy metal ions (Na, K, Ca, Hg, Cd, Pb)
	Organic compounds	Hydrocarbons, alcohol, organic solvents
	pH	H^+ ion
	Redox couples	Redox reactions
	Antibiotics	Fluoroquinolones, streptomycin, chloramphenicol
Biological	Proteins	Enzymes, biomarkers, antibodies
	Nucleic acids	DNA, RNA
	Hormones	Insulin, thyroid, cortisol
	Neurotransmitters	Dopamine, serotonin, glutamate, epinephrine, norepinephrine
	Cells and microorganisms	Cellular functions, pathogens (*Escherichia coli*, *Staphylococcus aureus*, etc.)
	Antigen and antibodies	SAR-CoV-2 antigen

belongs to this category and this type of sensors are mainly used to measure the pH changes.

- Conductometric sensors can be used to monitor electroactive and inactive species and it measures the specific conductance of an analyte species.
- Amperometric sensors measure the current response in order to determine the analyte concentration at a fixed potential applied to the cell. The electron transfer of the analyte is induced by the applied potential and the analyte concentration is determined by measuring the resultant current.

The different types of sensors are used in the analytical determination of a wide range of analytes. The important characteristics of electrochemical sensing are; in situ and real time monitoring of various analytes, higher sensitivity and selectivity, short response time, relatively low cost, portability, ease of automation, and miniaturization. The detection of target analyte is based on the redox reactions of analytes in the given electrochemical system. Electrochemical sensors play a pivotal role in exerting precise control over reactivity at the electrode/solution interface, rendering them indispensable in various domains such as electrocatalysis, electroanalysis, and corrosion mitigation. The applied potential stands out as a prominent variable for manipulating the physicochemical characteristics of this interface; however, its capacity for inducing significant changes and ensuring high selectivity remains limited. Consequently, a substantial portion of research efforts is devoted to the deliberate alteration of electrode surfaces, with the objective of imposing and regulating their properties.

6.5.1 Mechanism of Electrochemical Sensing

Electrochemical sensing is a powerful technique used to detect and quantify various chemical species (e.g., ions, molecules) by measuring the electrical response associated with an electrochemical reaction occurring at an electrode surface. The basic mechanism of electrochemical sensing involves the interaction between the analyte (the substance being analysed) and the working electrode, which leads to a measurable electrical signal. The key components and steps involved in electrochemical sensing are as follows:

- Electrode setup
  a. Working electrode: This is the primary electrode where the electrochemical reaction takes place. It is typically made of a conductive material (e.g., platinum, gold, carbon) and is in direct contact with the analyte.
  b. Reference electrode: The reference electrode has a stable, known potential and serves as a reference point for measuring the potential at the working electrode. Common reference electrodes include the Ag/AgCl electrode, saturated calomel electrode (SCE) etc.
  c. Counter electrode: The counter electrode provides a path for the flow of current during the electrochemical reaction. It helps maintain a balanced charge during the process.
- Electrochemical reaction: When the analyte of interest comes into contact with the surface of working electrode, it can undergo various electrochemical processes, which typically fall into two categories:
  a. Oxidation-reduction reactions (Redox): In these reactions, the analyte either loses electrons (oxidation) or gain electrons (reduction). The movement of electrons is associated with changes in the working electrode's potential, generating a current that can be measured.
  b. Ion-selective electrodes: These sensors selectively respond to specific ions in the solution by allowing the ions to enter or exit the electrode, leading to a change in potential.
- Electrical measurements: The potential at the working electrode is measured relative to the reference electrode, and the resulting voltage or current is recorded. This information is then used to quantify the concentration of the analyte.
- Signal processing: The recorded electrical signal may need to be processed, amplified, or filtered to improve the signal-to-noise ratio. This step can involve using electronic components and circuitry.
- Data analysis: The recorded data is then analysed to determine the concentration of the analyte. Calibration curves, standard additions, or other analytical methods may be used to relate the signal to the analytes concentration.

Electrochemical sensors are widely used in various applications, such as environmental monitoring, biomedical diagnostics, food quality control, and industrial process control. They offer advantages like high sensitivity, specificity, and the ability to work in complex sample matrices. The choice of electrode materials, electrode potential, and sensor design can be tailored to specific applications, making electrochemical sensing a versatile and valuable analytical tool.

Basically, the electrode/solution interface of an electrochemical sensor consists of a base electrode (substrate) modified with nano or functional materials. The surface modification of conventional electrodes is found to enhance the sensitivity and specificity of electrochemical sensor towards specific analytes. This class of electrodes are generally termed as chemically modified electrodes (CME) which provide more number of electrocatalytic active sites, and improved conductivity to the conventional unmodified electrodes. The common base electrodes are glassy carbon, platinum, graphite rod, gold, boron doped diamond electrode (BDDE), carbon paste, etc. which are combined with functional materials such as carbon-based nanostructures, metallic nanomaterials, hybrid materials, and polymers. In recent years, there has been a significant growth in the range of materials used for electrode modifiers. The surface modification of electrodes in electrochemical sensing is important especially in two instances; (i) when the interfering species generate signals that sum up or cover the signal of target analyte, the electroanalysis become inaccurate, (ii) in the simultaneous sensing of analytes with close redox potentials, the signal overlap can leads to less accurate result while using unmodified electrodes. In order to overcome these limitations, construction of electrochemical interfaces with materials of superior physico-chemical properties are of great interest.

6.5.2 MOFs as Electrode Modifiers

In electrochemical sensors, MOFs can be used as the chemically selective layer due to their large specific surface area which provide more number of accessible electrocatalytic sites, tunable chemical characteristics, ordered structure, presence of highly dispersed open metal sites/coordinatively unsaturated sites, etc. However, the presence of organic ligands throughout the framework leads to poor electrical conductivity of pristine MOFs which act as a hindrance for electrochemical detection. In order to improve the electrical conductivity of MOF without losing its innate characteristics, pristine MOF can be converted into its electrochemically active derivatives such as porous carbon and metal/metal oxide/metal sulphide or its composites by thermal and chemical treatments. In other words, MOF derivatives possess high surface area and porosity which provide more number of sites for electrocatalytic reactions along with the enhanced electrical conductivity and stability suitable for electrochemical sensing. Therefore, in order to design electrodes with high sensitivity, stability and selectivity, synthesis of efficient catalytic materials is of paramount important which can be used for the simultaneous detection of different analytes.

6.5.3 MOF-Derived Carbon Materials

As potential self-templates, MOFs can be readily transformed into materials with micro/nanoarchitectures, such as carbon and their hybrids. More significantly, the composition and morphology of MOF derivatives can be adjusted due to the wide choice of MOF precursors and different pyrolysis conditions. Furthermore, porous carbon is a highly conductive material with large surface area, hierarchical porous structure and good stability which make them a potential candidate for sensing applications. The diversity in their pore structure make them unique from other conventional carbon materials like graphene and carbon nanotube. In addition, the higher stability of derived carbon in aqueous phase compared to MOF materials is a preferable characteristic for designing sensing platforms.

6.5.3.1 MOF-Derived Carbon: Representative Examples

The heterostructures of MOFs with other functional materials like carbon and metal nanostructures shows higher electrochemical activity due to the synergistic effect of combining materials. Gao et al. developed an MOF-derived core-shell ternary heterostructure, Cu@C@ZIF-8, by the pyrolysis of Cu-MOF@ZIF-8 core-shell structure at 500°C for 2 h in argon atmosphere at a heating rate of 5°C/min [2]. During pyrolysis, Cu-MOF get completely transformed into Cu@C and the ZIF-8 preserved its structure even after carbonization. The obtained hybrid, Cu@C@ZIF-8, was applied for the electrochemical sensing of nitrite and it shows a wide linear range of 0.1–300 µM with a limit of detection 0.033 µM. The synergistic effect arising from the combination of ZIF-8 with large surface area and Cu@C having high catalytic activity leads to the excellent electrocatalytic activity of the ternary composite. Here, the ZIF-8 shell with appropriate pore size can selectively detect the analytes and hence account for its higher selectivity towards nitrite. Hence, the morphology as well as structure of MOF-derived material has significant effect on its sensing performance.

The tunable composition of MOF-derived materials is another important factor that influence its electrochemical activity. Heteroatom doping is a facile method to design MOF-derived carbon composite with enhanced electrocatalytic efficiency. This heteroatom can be derived either from the organic ligand (containing heteroatoms) in the precursor MOF or by the addition of element species during or after carbonization. For example, Dong et al. prepared a heteroatom-doped porous carbon composite, namely 3D SnS@Co,N-C hierarchical hollow microrod, where the 2D SnS nanoplates are anchored vertically on the Co-MOF (containing 2-methylimidazole ligand)-derived Co, N co-doped carbon rod [3]. The annealing of SnS_2@Co-MOF composite at 700°C, 3 h in N_2 environment is accompanied by a phase transformation of 2D SnS_2 nanoplates to 3D SnS phase which yields SnS@Co,N-C. The obtained composite is efficiently used for the simultaneous electrochemical sensing of biomolecules like epinephrine and L-tyrosine with wide linear ranges of 0.01–300 and 0.01–250 µM with limit of detection (LOD) 3.3 and 3.1 nM, respectively. Heteroatom doping increased the electrical conductivity and the number of electrocatalytic sites in the resultant composite. Furthermore, the hierarchical hollow structures shortens the ion diffusion pathway and accelerate the rate of electrochemical reaction which further enhances the electrochemical sensing ability of MOF-derived carbon structures. This work shows that MOFs can be used as efficient templates for the formation of various hierarchical hollow structures directly by pyrolysis.

Hybridization of MOF-derived carbon with other novel materials having intriguing properties is a widely explored strategy for fabricating sensing platforms. Huang et al. developed a heterostructure combining MXene and MOF-derived nitrogen-doped

TABLE 6.2
Examples of electrochemical sensors based on MOF-derived porous carbon

Material	Precursor MOF	Pyrolysis condition	Electrochemical technique	Analyte	Linear range	Limit of detection	Ref.
Co@NC/MWCNT	ZIF-67	800°C, 2 h	SWSV	Cd^{2+}, Pb^{2+}	0.12–2.5 µM	4.5 and 4.9 nM	[5]
MIP-PB-PC-CNT	MOF-5	1000°C, 2 h	DPV	Cysteine	1×10^{-13} to 1×10^{-7} M	6×10^{-15} M	[6]
NGR-NPC	ZIF-8	900°C, 3 h	DPV	AA, DA, UA	0.01–15.0 mM 0.08–350.0 µM 0.5–100.0 µM	1.99, 0.011 and 0.088 µM	[7]
ZIF-8C@rGO	ZIF-8	900°C, 5 h	DPV	HQ, CT	0.5–70 µM	0.073 and 0.076 µM	[8]
NPC-FJU-40-H	FJU-40-R	950°C, 2 h	DPV	HQ, CT	1–70, 1–100 µM	0.18, 0.31 µM	[9]
C-ZIF-67/PAN-800	ZIF-67	800°C, 3 h	DPV	HQ, CT	1–120, 1–200 µM	1.0 µM	[10]
EPC	IRMOF-8	1000°C, 5 h	SWV	CAP	1×10^{-8} to 1×10^{-6} M and 1×10^{-6} to 6×10^{-6} M	2.9×10^{-9} M	[11]

SWSV, square wave stripping voltammetry; NPC, N-doped porous carbon; NGR, N-doped graphene; FJU-40-R, $Zn(Trz)(R\text{-}BDC)_{1/2}$ (where, R = H or NH_2; Trz = 1,2,4-Triazole); EPC, exfoliated porous carbon; CAP, chloramphenicol.

porous carbon, termed as N-PC/alk-Ti_3C_2, for the electrochemical sensing of hydro-quinone and catechol [4]. The excellent electrical conductivity of alk-Ti_3C_2 combined with higher surface area of N-PC together make it a suitable sensing platform for phenolic compounds. The sensor shows a wide linear range of 0.5–150 µM with limit of detection 4.8 nM which make it suitable for its application in environmental analysis. Here, the ligand in MOF-5-NH_2 act as a nitrogen source to form C-N bond in nitrogen doped porous carbon, which forms hydrogen bond with the –OH functional groups in alk-Ti_3C_2. This hydrogen bonding favours the interaction between N-PC and alk-Ti_3C_2 and prevents the restacking of 2D Ti_3C_2. Furthermore, the heteroatom (N) present in the MOF-derived carbon matrix promotes the electron transfer and thereby increases the electrical conductivity (Table 6.2).

6.5.4 MOF-Derived Metal Species: Representative Examples

MOF-derived materials featuring a well-organized structure on conductive substrates, offer a promising solution to overcome the common limitation in many MOFs. This limitation involves agglomeration, resulting in a reduced number of accessible active sites for catalytic reactions. Addressing this issue, there has been significant interest in the design of MOF-derived materials characterized by ordered heterostructures. In one example, Ding et al. fabricated a binary nanocomposite with core-shell structure based on MOF-derived metal oxide for the electrochemical detection of glucose [12]. The self-supporting sensor consisting of $CuO_x@Co_3O_4$ where the CuO_x nanowire core is wrapped with Co_3O_4 nanoparticles shell on a conductive substrate copper (Cu) foam which further used to fabricate the electrode for sensing. The modified electrode exhibited a wide linear range of 0.1–1300 µM and limit of detection 36 nM for glucose oxidation as observed from CV studies. The synergistic catalytic effect between the metal oxide layers in a core-shell structure can accelerate the electrochemical activity of the material.

Besides core-shell structure, MOF-derived metal oxides exhibit different morphologies and structures as evident from the SEM and XRD analysis. For example, Zhang et al. prepared a hierarchical porous bimetallic oxide with micro-rice like structure, $ZnCo_2O_4$, by the pyrolysis (400°C, 2 h, air) of ZnCo-MOF and applied for the electrochemical sensing of glucose [13]. Here, the mesoporous structure of $ZnCo_2O_4$ favours the diffusion and adsorption of ions and molecules during electrochemical reaction. The amperometric technique shows that the sensor exhibited a wide linear range of 0.01–0.55 mM and limit of detection 5 µM.

MOF-derived transition metal sulphides and phosphides are another important category which possess porous structure with higher electrocatalytic activity. Zhang et al. prepared spindle-like Co-CuS microparticles using a bimetallic precursor, Co-Cu MOF, with the addition of thioacetamide (TAA) by heating in an oven at 110°C for 12 h. The synthesized Co-CuS was applied for the electrochemical sensing of glucose and the amperometric study shows the sensor exhibited a linear range of 0.001–3.6 mM with a detection limit of 0.1 µM. In another example, Wang and his co-workers developed porous metal phosphide by the calcination of in situ grown Ni-MOF-74 on graphene substrate in presence of sodium hypophosphite [14]. The calcination process performed at 275°C for 5 h in Ar atmosphere yields Ni_2P/graphene

TABLE 6.3
Examples of electrochemical sensors based on MOF-derived metal/metal oxide

Material	Precursor MOF	Pyrolysis condition	Electrochemical technique	Analyte	Linear range	LOD	Ref.
$NiCo_2O_4$	ZIF-67	350°C, 1 h	Amperometry	Glucose	0.18 µM– 5.1 mM	27 nM	[15]
$CuO@Co_3O_4$	ZIF-67	350°C, 3 h, air	Amperometry	Glucose	0.50 µM– 0.1 mM	0.23 µM	[16]
NiO@ZnO	Ni-Zn MOF	800°C, 2 h	DPV	Isoniazid	0.80–800 µM	0.25 µM	[17]
$CuO/Cu_2O@$ CuO/Cu_2O	$Cu(OH)_2@$ HKUST-1	350°C, 2 h, air	CV	Glucose	0.99–1330 µM	0.48 µM	[18]

(Ni_2P/G) where the Ni_2P nanoparticles are uniformly dispersed in a 3D framework wrapped with graphene. The composite was applied for the non-enzymatic glucose sensing and the Ni_2P/G modified GCE shows superior sensing performance with a wide linear range and lower detection limit (Table 6.3).

6.5.5 MOF-Derived Metal-Carbon Composite: Representative Examples

In most circumstances, the calcination of MOF can result in the formation of metal species nanoparticles homogeneously dispersed or embedded in the highly ordered porous carbon matrix resulting in the formation of MOF-derived metal-carbon composite. The composite possess advantages of both the components and overcomes the limitation with individual components for different applications. The metal/metal oxide nanoparticles with ultrasmall size, higher electrical conductivity combined with large specific surface area, porosity and stability in aqueous phase of porous carbon give rise to derived composites with excellent properties. In some cases, when the size of metal nanoparticles decrease, there is higher chance of agglomeration due to the increased surface energy of small nanoparticles. In order to overcome the agglomeration of particles, they can be immobilized on suitable support such as carbon nanostructures (graphene, carbon nanotube, carbon nanowire, meso/macro porous carbon). However, all these composite preparation require multi-step process, time consuming and involve complex procedure. Furthermore, owing to the consistent spacing between metal ions within the framework of MOF, when subjected to appropriate pyrolysis conditions, the organic ligand undergoes transformation into a porous carbon substance. Simultaneously, the metal ions are converted in situ into individual metal or metal oxide particles, and they are evenly distributed throughout the resulting carbon material. In essence, the utilization of MOFs as a precursor enables the one-step pyrolysis synthesis of uniformly dispersed metal compound nanoparticles distributed in porous carbon. Taking this into consideration, there are a number of research efforts focusing on the synthesis of such composites and its application for sensing.

TABLE 6.4
Examples of electrochemical sensors based on MOF-derived carbon-metal composites

Material	Precursor MOF	Pyrolysis condition	Electrochemical technique	Analyte	Linear range	LOD	Ref.
N/NiCu@C	N/Ni-Cu/ MOF	800°C, 2 h	CV	Neonicotinoid	0.5–60, 1–60, and 0.5–60 µM	0.017, 0.007 and 0.001 µM	[21]
NH$_2$-MIL-101(Fe)/ CNF	NH$_2$-MIL-101(Fe)	1000°C, 5 h	EIS	Tetracycline	0.1–105 nM	0.01 nM	[22]
CoN-CR	Co-ZIF-L	800°C, 2 h	Amperometry	Nitrite	0.5–4000 and 4000–8000 µM	0.17 µM	[23]

CoN-CR, Co@N-doped carbon rod.

For example, Wang et al. reported the synthesis of a binary mixed carbon/metal oxide material derived from MOF by the one-step pyrolysis of MIL series MOFs [19]. Notably, TiO$_2$/C$_{900}$ obtained by the pyrolysis of MIL-125(Ti) precursor exhibited the higher surface area and hierarchical micro-mesoporous characteristics useful for its application in the simultaneous detection of dihydroxybenzene isomers, i.e., hydroquinone (HQ) and catechol (CC). The study shows that the pyrolysis temperature and the nature of metal centres affect the internal structure and composition of obtained composite materials which in turn influence its properties. In another study, Li et al. synthesized Ni$_2$P/C by the calcination of Ni-MOF-74 with the addition of red phosphorous during carbonization [20]. The composite possess higher electrochemical activity and further employed as a glassy carbon modifier for the sensitive determination of uric acid. Under the optimum conditions, the sensor shows a linear range of 0–6.0 µM with limit of detection 7×10^{-8} M attributed to the higher charge transfer efficiency. The Ni$_2$P nanoparticles embedded in the porous carbon greatly increase the electronic conductivity (Table 6.4).

6.6 MOF-DERIVED SINGLE ATOM CATALYSTS

Single atom catalysts (SACs) are an emerging class of catalytic materials first reported in 2011. They are atomically dispersed metal species which are anchored on a suitable support such as porous carbon. The precise uniform distribution of single metal active sites allows for the easy regulation of catalytic activity and hence started widely exploring in electrocatalysis. However, the tendency of metal aggregation due to the high surface energy in the atomic level is a challenge that needs to be addressed while preparing SACs since it can lead to inferior catalytic activity. In order to avoid the aggregation of metal species, SACs need to be anchored on a suitable support that possess a porous structure which can accommodate uniform metal active sites. In this context, the use of MOF-based materials as template is a promising approach to prepare stable SACs with large surface area and porosity which possess increased number of accessible active sites favouring catalytic reaction. SACs derived from MOFs can be prepared in two ways: (1) by the direct pyrolysis of precursor MOF and (2) by introducing a second metal into the metal node or pores of MOF followed by its pyrolysis. The direct pyrolysis method mostly require acid treatment to remove the metal NPs formed along with SACs which makes the process complex and often leads to low metal loading. In the second method, the larger spatial separation between the two metals as well as the proper pyrolysis conditions allows for the removal of lower boiling point metal and results in the atomic dispersion of targeted metal on the porous carbon substrate.

SAC possess the advantages of porous structure combined with uniformly distributed active sites which in turn make it a potential candidate for electrocatalytic applications. For example, Liang et al. synthesized Co-single atomic site catalyst anchored on ZIF-derived N-doped porous carbon and applied it for the electrochemical sensing of levodopa (LD). Co-SAC with hierarchical porous structure derived from Zn-Co@ZIF-67 core-shell precursor has a uniform Co atom distribution with maximum atom utilization which enhances the electron transfer for sensing application. Atomically dispersed Co single atoms over highly conductive graphitic carbon layers provide an excellent sensing platform for LD detection [24]. In a recent study, Yang et al. prepared a Fe-SAC on nitrogen-doped carbon (Fe SAs/N-C) by the pyrolysis of NiFe LDH/ZIF-8 composite for the electrochemical detection of caffeic acid [25]. The Fe-N_4 active sites dispersed on N-C greatly enhanced the sensing performance of material. Here, the NiFe LDH act as the metal source and ZIF-8 provide the N-doped porous carbon support for the single atoms during pyrolysis of the composite. The area of SAC derived from MOF is still expanding with its applications in a wide range of electrochemical fields.

6.7 COLOURIMETRIC SENSING

Colourimetric sensors represent a significant subset of optical sensors, characterized by their ability to undergo a colour change upon interaction with an analyte which can be often detected using naked eye. In general, the principle of colourimetric sensing is based on the concept that specific interactions between an analyte and a reagent result in a discernible change in the colour of a solution or material. This change in colour is directly related to the concentration or presence of the analyte and can be observed with the naked eye or measured using UV-visible spectrophotometer. The fundamental steps involved in colourimetric sensing are as follows:

- Reagent-analyte interaction: A specific reagent is selected or designed to interact with the target analyte. This interaction can involve chemical reactions, complex formation, or binding events. The reagent is chosen to produce a colour change when it interacts with the analyte.
- Colour change: When the reagent interacts with the analyte, it undergoes a chemical change or physical alteration that leads to a change in its optical

properties, particularly its absorbance or transmission spectrum. This change results in a noticeable shift in colour, which is often evident as a change from one visible colour to another.

- Detection: The colour change can be observed visually by the human eye. For quantitative analysis or when the colour change is subtle, specialized colourimetric instruments, such as spectrophotometers or colorimeters, can be used to measure the intensity of light at specific wavelengths within the visible spectrum.
- Calibration: To determine the analyte concentration accurately, a calibration curve or standard reference samples are often employed. These standards provide a relationship between the colour change and the analyte concentration.
- Quantification: The extent of the colour change or the intensity of the colour, typically measured at a specific wavelength, is correlated with the analyte concentration. This information can be used to determine the concentration of the analyte in the sample.

Colourimetric sensing offers several advantages such as low cost, ease of operation, fast response and applicability in large detection volume making it a suitable technique in research areas such as biomedical and agriculture. In order to fabricate colourimetric aptasensors for the detection of various organic molecules MOFs containing transition metals are an excellent choice since they possess atomically dispersed metal nodes, ordered porosity and exhibit enzyme like activity. These properties can result in providing a micro-environment suitable for the substrates to interact with the active sites in MOF and hence increase the catalytic activity. However, the poor stability as well as the change in crystallinity and porosity of MOF due to external environment necessitate the development of stable MOF-derived materials for colourimetric sensing.

In general, MOF-derived materials can act as artificial enzymes (or nanozymes – 'nanomaterials with enzyme-like characteristics') and can be functionalized with suitable functional groups to which the aptamer is modified in order to fabricate the sensing platform for colourimetric method. Nanozymes are termed as 'artificial enzymes' which are nanomaterials with distinct catalytic activity and are characterized by high stability and low cost and. MOFs and their derivatives can mimic the properties of natural enzymes due to the large number of accessible active sites. In recent years, MOF-based nanozymes have gained considerable research attention and exhibits great application potential for the fast and onsite analysis of various analytes using simple instrumentation. In one example, Sun et al. presented a novel colourimetric aptasensor based on Fe doped ZIF-derived materials for the sensitive detection of organophosphorous pesticides. Here, the Fe-N-C nanozymes obtained by the pyrolysis of Fe-doped-ZIF was combined with the aptamer to construct sensing probe [26]. MOF-derived nanozymes possess large number of active sites, different compositions, tunable porosity and can be designed in test strip form which make it suitable for rapid and in situ monitoring of analytes (Figure 6.3 and Table 6.5).

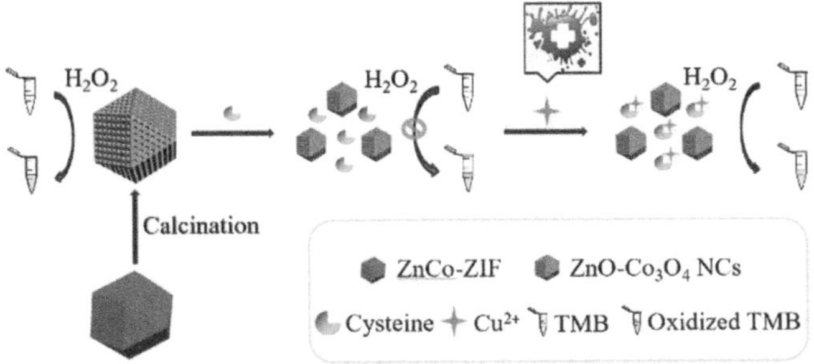

FIGURE 6.3 Schematic diagram of the colourimetric sensor for Cu^{2+} detection using ZnO–Co_3O_4 NCs. Reproduced with permission from Ref. [27].

TABLE 6.5
Examples of colourimetric sensors based on MOF-derived materials

Material	Precursor MOF	Calcination conditions	Type of nanozyme activity	Analyte	Linear range	LOD	Ref.
$MnCo_2O_4$	Co/Mn-MOF-74	300°C, 6 h, air	Peroxidase-like	Phenol	0.5–500 μM	0.2 μM	[28]
Cu NPs@C	Cu-MOF	400°C, 2 h, N_2	Peroxidase-like	Ascorbic acid	10 μM–1 mM	1.41 μM	[29]
Co-N/C-900	Co-Zn ZIF	900°C, 2 h, Ar	Peroxidase-like	L-Cysteine	1–40, 90–140 μM	33 nM	[30]
Co_3O_4/ cellulose	Co-MOF-74	300°C, 2 h, air	Peroxidase-like	Phenol	0.5–300 μM	1.02 μM	[31]
Co NPs/MC	NH_2-MIL-88(Fe)	500°C, 2 h, N_2	Peroxidase-like	Glucose	0.25–30 μM	150 nM	[32]
Fe-N-C SAN	Fe-doped ZIF-8	900°C, 2 h, N_2	Peroxidase-like	BChE	0.1–10 U/L	0.054 U/L	[33]
ZnO-Co_3O_4	ZnCo-ZIF	350°C, 3 h, air	Peroxidase-like	Cu^{2+}	2–100 nM	1.08 nM	[27]
Co-N/ S-HCN	ZIF-67@TA	800°C, 1 h	Oxidase-like	Hg^{2+} Fe^{3+}	1–5, 5–9 μM	0.09 μM	[34]

SAN, single atom nanozyme; BChE, butyrylcholinesterase; TA, thiomalic acid.

6.8 MOF-DERIVED MATERIALS FOR GAS SENSING

Extensive exposure to airborne pollutants, including volatile organic compounds (VOCs – e.g., acetone, toluene, benzene, acetic acid), hydrogen sulphide (H_2S), formaldehyde, nitrogen oxides (NO_x), among others, is widely acknowledged to have adverse effects on human health. For instance, the excessive content of n-butanol in the atmosphere cause adverse health effects such as dizziness, headache and dermatitis whereas exposure to acetone can cause eye irritation, nausea and narcosis [35]. In addition, H_2S gas in air may affect the nervous system and when it exceeds 250 ppm, there is high chance of mortality [36]. These facts underscore the critical importance of understanding and mitigating the health risks associated with long-term exposure to gaseous pollutants.

Gas sensors are a broad category of analytical devices used to detect mainly toxic, flammable, and/or greenhouse gases. The sensor consists of a recognition element (sensing material) and a transducer. The recognition layer interact with the gas analyte molecules thereby causing a modulation in its electric, magnetic, dielectric, optical, gravimetric or colourimetric properties. The change in the properties of sensing element is converted into a measurable signal by the transducer. Based on the transduction mechanism, gas sensors are broadly classified into electrically transduced, optical and mass sensitive sensors. Among these, the electrically transduced gas sensors are the most explored with MOF-derived materials which is further subdivided into electrochemical, chemiresistive, field-effect transistor (FET) based, chemicapacitors, etc. (Figure 6.4).

The most studied materials for gas sensors include metal oxide semiconductors (TiO_2, ZnO, Fe_2O_3, SnO_2, NiO, and Co_3O_4), carbon-based materials (graphene, CNT, reduced graphene oxide) and polymer-based compounds (PANI, polypyrrole, polyacetylene, etc.) along with few reports of ionic liquids, 2D layered materials (MoS_2, WS_2) and biomaterials. Even though each of these materials possess several advantages, certain limitations such as poor selectivity/sensitivity, sluggish sensing kinetics, high energy requirement and complicated sensor configuration hinder its application for the fabrication of commercially viable high performance gas sensors.

In this circumstances, MOF-derived materials emerged as an effective choice for the design of gas sensors due to its intriguing physico-chemical and structural

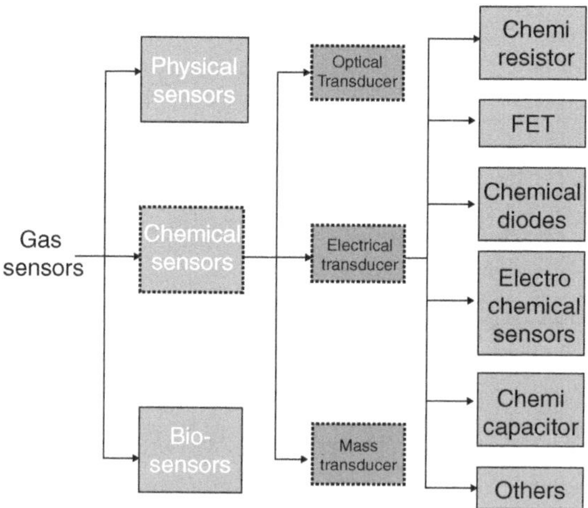

FIGURE 6.4 Classification of gas sensors. Reproduced with permission from Ref. [37].

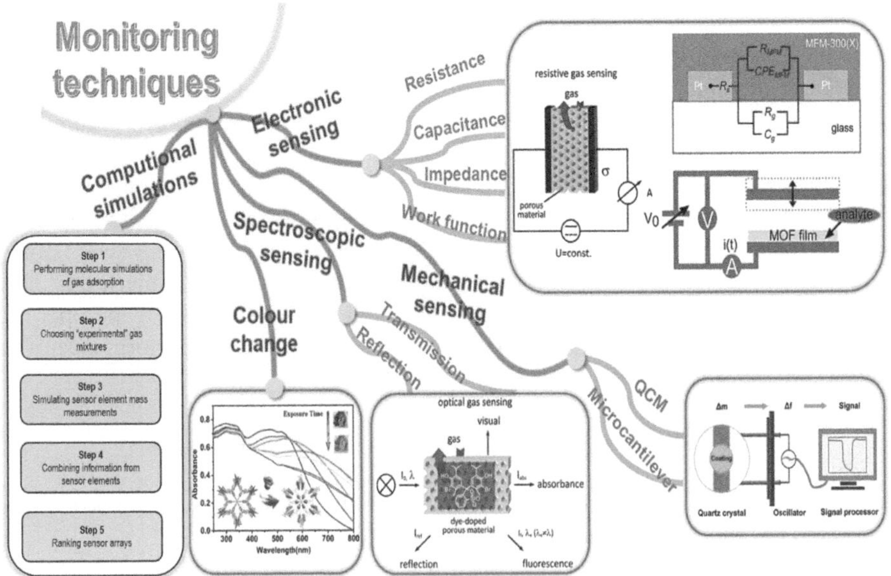

FIGURE 6.5 Schematic of the summary of sensing parameters in gas sensors. Reproduced with permission from Ref. [38].

features. In particular, the tenability of composition and structure of MOF results in diverse MOF-derived materials which opens the pathway for the design of novel materials for the sensing of specific gas molecules. Furthermore, the tunable structure of derived materials allows the selective permeation of gas molecules which leads to higher selectivity of sensor. The other characteristics of MOF derivatives that make them ideal as gas sensing materials are high chemical stability, resistance to temperature, less toxicity and large specific surface area. MOF derivatives undergoes a change in the physical and/or chemical properties due to the adsorption/desorption of target gas molecules on the surface of materials (Figure 6.5).

6.8.1 ELECTRICALLY TRANSDUCED GAS SENSORS

This section briefly discusses some important electrically transduced gas sensors designed using MOF-derived materials. Basically, this type of sensor consists of sensing layer and electrical transducer. When the gaseous analyte interact with the sensing layer, there may be changes in the electrical parameters of sensing layer such as conductivity, resistivity, and dielectric constant. These changes are converted into measurable electric signals such as current, potential, capacitance, and resistance/impedance. Typically, electrically transduced gas sensing involves three steps: diffusion of gas on the surface of sensing layer, the surface reaction (analyte–material interaction) and transfer of electrical parameters. Among the different types of electrically transduced gas sensors, chemiresistive sensors are the most explored one using MOF-derived materials.

6.8.2 CHEMIRESISTIVE GAS SENSORS

This type of sensors detect the composition and concentration of target gas by converting the chemical information into electrical signal, i.e., the variation in the resistance of sensor relative to the change in analyte concentration is measured. The change in resistance significantly depends on the operating temperature of the sensor. The sensing process involves the adsorption of gas molecules, reaction followed by desorption on the sensing surface. Above the optimum operating temperature, the sensing material starts to decompose which in turn cause a decrease in sensitivity. In addition, the sensing layer becomes more susceptible to other interfering gases at higher temperature resulting in a decrease in selectivity. In short, the operating temperature is an important sensing parameter for evaluating the performance of gas sensor. The sensor response (R) is defined as;

$$R = R_g/R_a \text{ for oxidizing gases; } R = R_a/R_g \text{ for reducing gases}$$

where R_g and R_a denote the electrical resistance of the gas sensor when exposed to air and given concentration of target gas, respectively.

For example, Zhang et al. developed a p-n heterojunction of $CoSe_2$@N-doped carbon decorated with MWCNT nanocomposite utilizing ZIF-67 template and in situ selenization of cobalt ions. The synergistic effect of nanocomposite results in sensor with superior characteristics for ammonia gas sensing. The sensor shows a response of 93.3% towards 10 ppm ammonia at room temperature and lower limit of detection (0.1 ppm) [39]. In another study, Jo et al. prepared In_2O_3 hollow spheres containing PdO nanoparticles encapsulated Co_3O_4 clusters and applied for the detection of breath acetone [40]. The Pd NPs@Co_3O_4 clusters were derived by the annealing of a 2-D precursor MOF, Co-ZIF-L, intercalated with Pd NPs at 500°C. The fabricated sensor, PdO/Co_3O_4-In_2O_3 film shows a high response of 145.9–5 ppm acetone concentration at an operating temperature of 225°C (Table 6.6).

6.9 CONCLUSION

In this chapter, the application of MOF-derived materials in different fields of sensing is discussed in detail with a focus on electrochemical, colourimetric, and chemiresistive gas sensing. This section intends to summarize the different types of MOF derivatives reported so far in the sensing of analytes of various interest. The large surface area and porous structure inherited from precursor MOF along with the enhanced electrical conductivity and stability makes the MOF-derived carbon and its hybrids an excellent candidate for electrochemical sensor development. SACs derived from MOF is a relatively new area and still needs to be explored more. Typically, SAC possess atomically dispersed active metal sites anchored on porous carbon and this exhibits a unique structure providing large number of accessible active sites for the catalytic process as well as the porous carbon support enhance the electrical conductivity of the prepared sensing materials. Furthermore, the application of MOF-derived metal oxides in the detection of gases which cause environmental pollution and health hazard, especially electrically transduced gas sensors, are in the developing stage in the recent years.

TABLE 6.6

Examples of chemiresistive gas sensors based on MOF-derived materials

Material	MOF precursor	Pyrolysis conditions	Target gas	Operating temp.	Response (R_g/R_a or R_a/R_g)	LOD	Ref.
ZnO	ZIF-8	300°C–600°C, air	NO_2 (1 ppm)	200°C	51.41		[41]
In_2O_3/g-C_3N_4	MIL-68(In)	450°C, 3 h	NO_x (100 ppm)	RT	294.9	0.0029 ppm	[42]
TiN@CuO	Cu-BTC	400°C, 2 h	H_2S (10 0 ppm)	50°C	34.6		[43]
Co_3O_4/$NiCo_2O_4$	ZIF-67	550°C, 2 h, air	H_2S (100 ppm)	50°C	57		[44]
CuO/In_2O_3	CPP-3(In)	450°C, 1 h	H_2S (5 ppm)	70°C	229.3	200 ppb	[45]
IGO@ZnO	MIL-68(In/Ga)	500°C, air	Acetone (1000 ppm)	300°C	164		[46]
PdO-Co_3O_4	ZIF-67	600°C, 1 h	Acetone (5 ppm)	350°C	2.51	0.1 ppm	[47]
ZnO/NiO	Zn/Ni-MOF	400°C, 2 h, air	n-Propanol (500 ppm)	275°C	280.2	200 ppb	[48]
Co_3O_4	Co_5-MOF	350°C	Formaldehyde	170°C		10 ppm	[49]

Still, there is ample space for the design and application of derived materials in gas sensors based on other transduction mechanisms such as optical and mass sensitive sensors. Also, the development of stable SACs from MOF without any metal NP aggregation by reducing the harsh treatment conditions after pyrolysis is a challenge in the materials development in electrochemical application. Finally, the incorporation of MOF derivatives in the fabrication of sensing devices that can withstand real-time analysis conditions by minimizing the matrix effect need to be explored in detail.

REFERENCES

1. Mandoj, F.; Nardis, S.; Di Natale, C.; Paolesse, R. Porphyrinoid Thin Films for Chemical Sensing. In *Encyclopedia of Interfacial Chemistry: Surface Science and Electrochemistry*; Elsevier, 2018, pp 422–443. www.sciencedirect.com/referencew ork/9780128098943/encyclopedia-of-interfacial-chemistry
2. Gao, F.; Tu, X.; Yu, Y.; Gao, Y.; Zou, J.; Liu, S.; Qu, F.; Li, M.; Lu, L. Core–Shell Cu@ C@ ZIF-8 Composite: A High-Performance Electrode Material for Electrochemical Sensing of Nitrite with High Selectivity and Sensitivity. *Nanotechnology*, 2022, *33* (22), 225501.
3. Dong, Y.; Li, J.; Zhang, L. 3D Hierarchical Hollow Microrod Via In-Situ Growth 2D SnS Nanoplates on MOF Derived Co, N Co-Doped Carbon Rod for Electrochemical Sensing. *Sens Actuators, B: Chem*, 2020, *303*, 127208.
4. Huang, R.; Liao, D.; Chen, S.; Yu, J.; Jiang, X. A Strategy for Effective Electrochemical Detection of Hydroquinone and Catechol: Decoration of Alkalization-Intercalated Ti3C2 with MOF-Derived N-Doped Porous Carbon. *Sens Actuators, B: Chem*, 2020, *320*, 128386.
5. Zhao, J.; Long, Y.; He, C.; Yang, H.; Zhao, S.; Luo, X.; Huo, D.; Hou, C. Simultaneous Electrochemical Detection of Cd2+ and Pb2+ Based on an MOF-Derived Carbon

Composite Linked with Multiwalled Carbon Nanotubes. *ACS Sustainable Chem Eng*, 2023, *11* (6), 2160–2171.

6. Duan, D.; Yang, H.; Ding, Y.; Li, L.; Ma, G. A Three-Dimensional Conductive Molecularly Imprinted Electrochemical Sensor Based on MOF Derived Porous Carbon/Carbon Nanotubes Composites and Prussian Blue Nanocubes Mediated Amplification for Chiral Analysis of Cysteine Enantiomers. *Electrochim Acta*, 2019, *302*, 137–144.

7. Luo, G.; Deng, Y.; Zhang, X.; Zou, R.; Sun, W.; Li, B.; Sun, B.; Wang, Y.; Li, G. A ZIF-8 Derived Nitrogen-Doped Porous Carbon and Nitrogen-Doped Graphene Nanocomposite Modified Electrode for Simultaneous Determination of Ascorbic Acid, Dopamine and Uric Acid. *New J Chem*, 2019, *43* (43), 16819–16828.

8. Chen, H.; Wu, X.; Lao, C.; Li, Y.; Yuan, Q.; Gan, W. MOF Derived Porous Carbon Modified rGO for Simultaneous Determination of Hydroquinone and Catechol. *J Electroanal Chem*, 2019, *835*, 254–261.

9. Zheng, S.; Wu, D.; Huang, L.; Zhang, M.; Ma, X.; Zhang, Z.; Xiang, S. Isomorphic MOF-Derived Porous Carbon Materials as Electrochemical Sensor for Simultaneous Determination of Hydroquinone and Catechol. *J Appl Electrochem*, 2019, *49*, 563–574.

10. Zhang, M.; Li, M.; Wu, W.; Chen, J.; Ma, X.; Zhang, Z.; Xiang, S. MOF/PAN Nanofiber-Derived N-Doped Porous Carbon Materials with Excellent Electrochemical Activity for the Simultaneous Determination of Catechol and Hydroquinone. *New J Chem*, 2019, *43* (9), 3913–3920.

11. Xiao, L.; Xu, R.; Yuan, Q.; Wang, F. Highly Sensitive Electrochemical Sensor for Chloramphenicol Based on MOF Derived Exfoliated Porous Carbon. *Talanta*, 2017, *167*, 39–43.

12. Ding, J.; Zhong, L.; Wang, X.; Chai, L.; Wang, Y.; Jiang, M.; Li, T.-T.; Hu, Y.; Qian, J.; Huang, S. General Approach to MOF-Derived Core-Shell Bimetallic Oxide Nanowires for Fast Response to Glucose Oxidation. *Sens Actuators, B: Chem*, 2020, *306*, 127551.

13. Zhang, D.; Wang, Z.; Li, J.; Hu, C.; Zhang, X.; Jiang, B.; Cao, Z.; Zhang, J.; Zhang, R. MOF-Derived ZnCo2O4 Porous Micro-Rice with Enhanced Electro-Catalytic Activity for the Oxygen Evolution Reaction and Glucose Oxidation. *RSC Adv*, 2020, *10* (15), 9063–9069.

14. Zhang, Y.; Xu, J.; Xia, J.; Zhang, F.; Wang, Z. MOF-Derived Porous Ni2P/Graphene Composites with Enhanced Electrochemical Properties for Sensitive Nonenzymatic Glucose Sensing. *ACS Appl Mater Interfaces*, 2018, *10* (45), 39151–39160.

15. Feng, Y.; Xiang, D.; Qiu, Y.; Li, L.; Li, Y.; Wu, K.; Zhu, L. MOF-Derived Spinel NiCo2O4 Hollow Nanocages for the Construction of Non-enzymatic Electrochemical Glucose Sensor. *Electroanalysis*, 2020, *32* (3), 571–580.

16. Muthurasu, A.; Kim, H. Y. Fabrication of Hierarchically Structured MOF-Co3O4 on Well-Aligned CuO Nanowire with an Enhanced Electrocatalytic Property. *Electroanalysis*, 2019, *31* (5), 966–974.

17. Wang, J.; Zhao, J.; Yang, J.; Cheng, J.; Tan, Y.; Feng, H.; Li, Y. An Electrochemical Sensor Based on MOF-Derived NiO@ ZnO Hollow Microspheres for Isoniazid Determination. *Microchim Acta*, 2020, *187*, 1–8.

18. Yu, C.; Cui, J.; Wang, Y.; Zheng, H.; Zhang, J.; Shu, X.; Liu, J.; Zhang, Y.; Wu, Y. Porous HKUST-1 Derived CuO/Cu2O Shell Wrapped Cu(OH)2 Derived CuO/Cu2O Core Nanowire Arrays for Electrochemical Nonenzymatic Glucose Sensors with Ultrahigh Sensitivity. *Appl Surf Sci*, 2018, *439*, 11–17.

19. Wang, Z.; Li, M.; Ye, Y.; Yang, Y.; Lu, Y.; Ma, X.; Zhang, Z.; Xiang, S. MOF-Derived Binary Mixed Carbon/Metal Oxide Porous Materials for Constructing Simultaneous Determination of Hydroquinone and Catechol Sensor. *J Solid State Electrochem*, 2019, *23*, 81–89.

20. Li, G.; Liu, S.; Liu, D.; Zhang, N. MOF-Derived Porous Nanostructured Ni2P/C Material with Highly Sensitive Electrochemical Sensor for Uric Acid. *Inorg Chem Commun*, 2021, *130*, 108713.

21. Zhangsun, H.; Wang, Q.; Xu, Z.; Wang, J.; Wang, X.; Zhao, Y.; Zhang, H.; Zhao, S.; Li, L.; Li, Z. NiCu Nanoalloy Embedded in N-Doped Porous Carbon Composite as Superior Electrochemical Sensor for Neonicotinoid Determination. *Food Chem*, 2022, *384*, 132607.

22. Song, J.; Huang, M.; Lin, X.; Li, S. F. Y.; Jiang, N.; Liu, Y.; Guo, H.; Li, Y. Novel Fe-Based Metal–Organic Framework (MOF) Modified Carbon Nanofiber as a Highly Selective and Sensitive Electrochemical Sensor for Tetracycline Detection. *Chem Eng J*, 2022, *427*, 130913.

23. Yang, Z.; Zhou, X.; Yin, Y.; Xue, H.; Fang, W. Metal-Organic Framework Derived Rod-Like Co@Carbon for Electrochemical Detection of Nitrite. *J Alloys Compd*, 2022, *911*, 164915.

24. Liang, W.; Gao, M.; Li, Y.; Tong, Y.; Ye, B.-C. Single-Atom Electrocatalysts Templated by MOF for Determination of Levodopa. *Talanta*, 2021, *225*, 122042.

25. Yang, X.; Lv, S.; Gan, L.; Wang, C.; Wang, Z.; Zhang, Z. Single-Fe-Atom Catalyst for Sensitive Electrochemical Detection of Caffeic Acid. *ACS Appl Mater Interfaces*, 2023, *15*, 53189–53197.

26. Shen, Z.; Xu, D.; Wang, G.; Geng, L.; Xu, R.; Wang, G.; Guo, Y.; Sun, X. Novel Colorimetric Aptasensor Based on MOF-Derived Materials and Its Applications for Organophosphorus Pesticides Determination. *J Hazard Mater*, 2022, *440*, 129707.

27. Lv, J.; Zhang, C.; Wang, S.; Li, M.; Guo, W. MOF-Derived Porous ZnO-Co3O4 Nanocages as Peroxidase Mimics for Colorimetric Detection of Copper (ii) Ions in Serum. *Analyst*, 2021, *146* (2), 605–611.

28. Zhang, K.; Lu, L.; Liu, Z.; Cao, X.; Lv, L.; Xia, J.; Wang, Z. Metal-Organic Frameworks-Derived Bimetallic Oxide Composite Nanozyme Fiber Membrane and the Application to Colorimetric Detection of Phenol. *Colloids Surf, A*, 2022, *650*, 129662.

29. Tan, H.; Ma, C.; Gao, L.; Li, Q.; Song, Y.; Xu, F.; Wang, T.; Wang, L. Metal–Organic Framework-Derived Copper Nanoparticle@ Carbon Nanocomposites as Peroxidase Mimics for Colorimetric Sensing of Ascorbic Acid. *Chem Eur J*, 2014, *20* (49), 16377–16383.

30. Chen, J.; Hu, S.; Cai, Y.; Liu, X.; Wu, Y.; Dai, Y.; Wang, Z. Co–N/C-900 Metal–Organic Framework-Derived Nanozyme as a H2O2-Free Oxidase Mimic for the Colorimetric Sensing of l-Cysteine. *Analyst*, 2022, *147* (5), 915–922.

31. Hou, C.; Fu, L.; Wang, Y.; Chen, W.; Chen, F.; Zhang, S.; Wang, J. Co-MOF-74 Based Co3O4/Cellulose Derivative Membrane as Dual-Functional Catalyst for Colorimetric Detection and Degradation of Phenol. *Carbohydr Polym*, 2021, *273*, 118548.

32. Dong, W.; Zhuang, Y.; Li, S.; Zhang, X.; Chai, H.; Huang, Y. High Peroxidase-Like Activity of Metallic Cobalt Nanoparticles Encapsulated in Metal–Organic Frameworks Derived Carbon for Biosensing. *Sens Actuators, B*, 2018, *255*, 2050–2057.

33. Niu, X.; Shi, Q.; Zhu, W.; Liu, D.; Tian, H.; Fu, S.; Cheng, N.; Li, S.; Smith, J. N.; Du, D. Unprecedented Peroxidase-Mimicking Activity of Single-Atom Nanozyme with Atomically Dispersed Fe–Nx Moieties Hosted by MOF Derived Porous Carbon. *Biosens Bioelectron*, 2019, *142*, 111495.

34. Yang, X.; Feng, M.; Zhang, X.; Huang, Y., Co, N. S Co-doped Hollow Carbon with Efficient Oxidase-Like Activity for the Detection of Hg2+ and Fe3+ Ions. *Microchem J*, 2023, *187*, 108383.

35. (a) Chen, Y.; Shen, Z.; Jia, Q.; Zhao, J.; Zhao, Z.; Ji, H. A CuO–ZnO Nanostructured p-n Junction Sensor for Enhanced N-Butanol Detection. *RSC Adv*, 2016, *6* (3),

2504–2511; (b) Drmosh, Q. A.; Olanrewaju Alade, I.; Qamar, M.; Akbar, S. Zinc Oxide-Based Acetone Gas Sensors for Breath Analysis: A Review. *Chem Asian J,* 2021, *16* (12), 1519–1538.

36. Shaik, R.; Kampara, R. K.; Kumar, A.; Sharma, C. S.; Kumar, M. Metal Oxide Nanofibers Based Chemiresistive H2S Gas Sensors. *Coord Chem Rev,* 2022, *471,* 214752.

37. Yao, M.-S.; Li, W.-H.; Xu, G. Metal–Organic Frameworks and Their Derivatives for Electrically-Transduced Gas Sensors. *Coord Chem Rev,* 2021, *426,* 213479.

38. Zhang, R.; Lu, L.; Chang, Y.; Liu, M. Gas Sensing Based on Metal-Organic Frameworks: Concepts, Functions, and Developments. *J Hazard Mater,* 2022, *429,* 128321.

39. Mi, Q.; Zhang, D.; Zhang, X.; Wang, D. Highly Sensitive Ammonia Gas Sensor Based on Metal-Organic Frameworks-Derived CoSe2@ Nitrogen-Doped Amorphous Carbon Decorated with Multi-Walled Carbon Nanotubes. *J Alloys Compd,* 2021, *860,* 158252.

40. Jo, Y.-M.; Lim, K.; Choi, H. J.; Yoon, J. W.; Kim, S. Y.; Lee, J.-H. 2D Metal-Organic Framework Derived Co-loading of Co3O4 and PdO Nanocatalysts on In2O3 Hollow Spheres for Tailored Design of High-Performance Breath Acetone Sensors. *Sens Actuators, B,* 2020, *325,* 128821.

41. Ren, X.; Xu, Z.; Liu, D.; Li, Y.; Zhang, Z.; Tang, Z. Conductometric NO2 Gas Sensors Based on MOF-Derived Porous ZnO Nanoparticles. *Sens Actuators, B,* 2022, *357,* 131384.

42. Liu, Y.; Liu, J.; Pan, Q.; Pan, K.; Zhang, G. Metal-Organic Framework (MOF) Derived In2O3 and g-C3N4 Composite for Superior NOx Gas-Sensing Performance at Room Temperature. *Sens Actuators, B,* 2022, *352,* 131001.

43. Amu-Darko, J. N. O.; Hussain, S.; Zhang, X.; Ouladsmane, M.; Issaka, E.; Ali, S.; Wang, M.; Qiao, G. Exploring the Gas-Sensing Properties of MOF-Derived TiN@ CuO as a Hydrogen Sulfide Sensor. *Chemosphere,* 2023, *337,* 139401.

44. Tan, J.; Hussain, S.; Ge, C.; Wang, M.; Shah, S.; Liu, G.; Qiao, G. ZIF-67 MOF-Derived Unique Double-Shelled Co3O4/NiCo2O4 Nanocages for Superior Gas-Sensing Performances. *Sens Actuators, B,* 2020, *303,* 127251.

45. Li, S.; Xie, L.; He, M.; Hu, X.; Luo, G.; Chen, C.; Zhu, Z. Metal-Organic Frameworks-Derived Bamboo-Like CuO/In2O3 Heterostructure for High-Performance H2S Gas Sensor with Low Operating Temperature. *Sens Actuators, B,* 2020, *310,* 127828.

46. Zhang, Y.; Jia, C.; Kong, Q.; Fan, N.; Chen, G.; Guan, H.; Dong, C. ZnO-Decorated In/Ga Oxide Nanotubes Derived from Bimetallic In/Ga MOFs for Fast Acetone Detection with High Sensitivity and Selectivity. *ACS Appl Mater Interfaces,* 2020, *12* (23), 26161–26169.

47. Koo, W.-T.; Yu, S.; Choi, S.-J.; Jang, J.-S.; Cheong, J. Y.; Kim, I.-D. Nanoscale PdO Catalyst Functionalized Co3O4 Hollow Nanocages Using MOF Templates for Selective Detection of Acetone Molecules in Exhaled Breath. *ACS Appl Mater Interfaces,* 2017, *9* (9), 8201–8210.

48. Zhao, Y.; Wang, S.; Zhai, X.; Shao, L.; Bai, X.; Liu, Y.; Wang, T.; Li, Y.; Zhang, L.; Fan, F. Construction of Zn/Ni Bimetallic Organic Framework Derived ZnO/NiO Heterostructure with Superior N-Propanol Sensing Performance. *ACS Appl Mater Interfaces,* 2021, *13* (7), 9206–9215.

49. Zhou, W.; Wu, Y.-P.; Zhao, J.; Dong, W.-W.; Qiao, X.-Q.; Hou, D.-F.; Bu, X.; Li, D.-S. Efficient Gas-Sensing for Formaldehyde with 3D Hierarchical Co3O4 Derived from Co5-Based MOF Microcrystals. *Inorg Chem,* 2017, *56* (22), 14111–14117.

7 Energy Applications

7.1 INTRODUCTION

Considering the ever-increasing energy consumption, focusing on alternative energy sources other than fossil fuels is essential. Only about 29% of global electricity is generated with renewables, and the number varies widely concerning countries. Moreover, the increase in adaptation of renewable energy sources is considerably low, from 11% to 29% between 2011 and 2022. There is a considerable need to research renewable technologies to reach net-zero energy consumption eventually (Figure 7.1). Solar cells are among the primarily researched alternative energy sources owing to the abundantly available solar energy. Various rechargeable batteries and supercapacitors are continuously improving for better efficiency and lower carbon print. MOFs and MOF-derived materials play an essential role in the advancements of many such energy-related applications. Owing to their porosity, stability, and low costs, MOFs are used in various forms to obtain MOF-derived materials, which, in turn, find their place in numerous energy-based materials. MOF precursors make excellent candidates for energy-related applications due to their tunability, diverse pore channels, large specific surface areas and good stability. This chapter overviews various critical energy applications being revolutionized by MOF-derived components.

7.2 SOLAR CELL

Solar cells work under the photovoltaic effect, in which the solar energy is converted into electrical energy. Some common types of solar cells include the conventional P-N Junction device, plasmonic solar cell, dye-sensitized solar cell, and organic and perovskite solar cells. In the traditional junction of p-n solar cells, silicon layers incorporating molecular dopants [1–3] are used as semiconducting material. In Plasmonic solar cells, a layer of metal nanoparticles is used in conjunction with silicon layers, which increases efficiency. The metal nanoparticle layer exhibits plasmon resonance, by which the effective path length of the incident light is increased 10–100 folds [4]. This enhances the adsorption of solar energy and decreases the required silicon layer thickness (and so the overall device costs). In dye-sensitized solar cells (DSSCs), the semiconductor is only used for charge transport, whereas the semiconductor in the PN junction is responsible for light absorption, carrier separation and transportation.

DOI: 10.1201/9781003432357-9

FIGURE 7.1   The transition from fossil fuels.

The photoelectrons are generated with the help of separate photosensitive dyes in the case of DSSCs. The photoelectron generation is achieved using organic active layers in the case of organic solar cells and perovskite interfacial layers in the case of perovskite solar cells. This section explores the application of MOF-derived materials in DSSCs and organic and perovskite solar cells.

7.2.1   MOF-Derived Materials in DSSCs

Out of the most studied solar cells, dye-sensitized solar cells are deemed the most cost-effective [5]. Although less efficient than conventional solar cells, their flexibility, weight, and costs make them attractive. DSSCs have attracted researchers due to their low cost and high efficiency [6]. The first DSSC was reported by Grätzel in 1991, and since then, this field has gathered massive interest among researchers [7]. MOFs and MOF-derived materials are increasingly used in DSSCs to improve various properties.

The dye-sensitized solar cells (DSSCs) are the ones where the design principles are quite different to those of the conventional PN junction cells. The construction and working principle of DSSCs is illustrated in Figure 7.2. The various working components of DSSCs are photoanode, sensitizer, electrode and electrolyte. Notably, the function of light adsorption and charge carrier transport is separated in DSSCs, which makes them distinct from the conventional PN junction solar cells. The light incident is absorbed by the photosensitizer coated over the photoanode. The electrons are absorbed due to the photon absorption. The excited electrons get

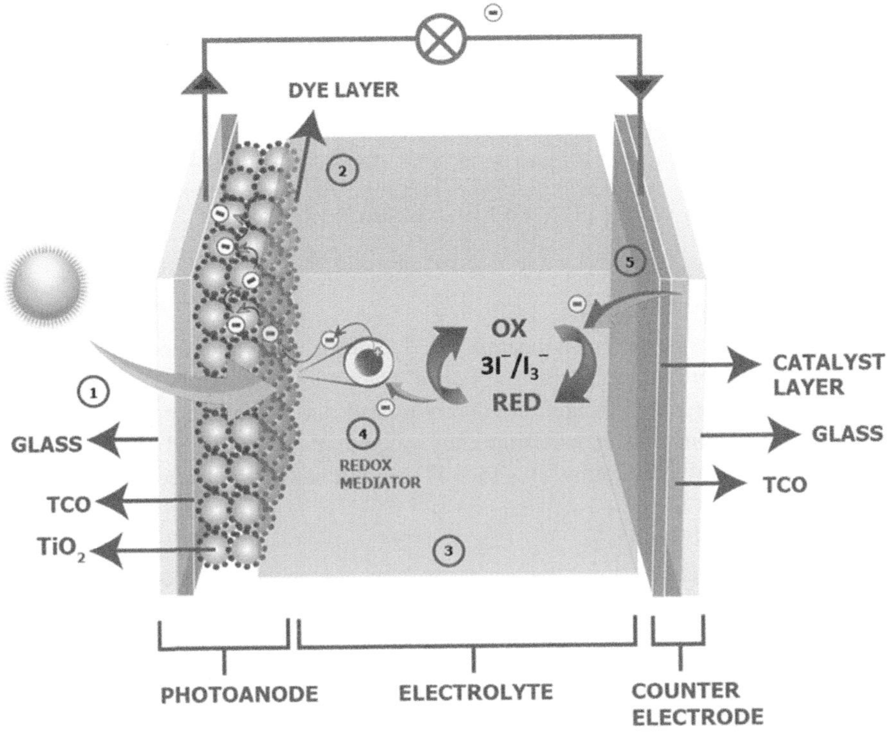

FIGURE 7.2   Working of a DSSC. Reprinted from [8] with open access license.

injected into the nanoporous anode. The injected electrons get diffused through the back contact (if present) and pass an external circuit to reach the electrode. The electrons flow from the electrode to the electrolyte and back to the dye, initiating a series of electrochemical reactions and reducing the dye back to its ground state. DSSCs are comprised of photosensitizers, photoanodes, counter electrodes, and electrolytes.

7.2.1.1   Photosensitizer

The photosensitizer is where the light incident occurs first within the solar cell setup. The photosensitizer is added as an additional component over the photoanode. This is a critical component as it produces photoexcited electrons for the DSSC to function. The photosensitizers absorb the solar photons, and the photoexcited electrons are transferred to the photoanode on which it sits. Some requirements for a good photosensitizer include appropriate LUMO and highest occupied molecular orbital (HOMO) levels for charge injection, photostability, adhesion to the semiconductor surface, and high molar extinction coefficients. Some common sensitizers used are organic sensitizers [9, 10], Ruthenium based sensitizers [11–13], metal-free sensitizers [14, 15], porphyrin sensitizers [16, 17], phenolic dyes [18, 19], water-soluble dyes [20, 21], quantum-dot sensitizers [22].

7.2.1.2 Photoanode

Photoanode (also called 'working electrode') is where the electrons from photosensitizers are carried to the external circuit. Hence, it acts as a transport and structural agent for the photosensitizer. Ideal photoanode should have low electrical resistance, fast charge transport, high surface area, and good optical transmittance [23]. Some common photoanode materials used are titanium dioxide [24, 25], FTO (fluorine-doped tin oxide) [26], ITO (indium-doped tin oxide) [27] and zinc oxide [28, 29].

7.2.1.3 Counter Electrode

The counter electrode (referred to simply as 'electrode') is where the electrons reduce the redox mediator in the electrolyte solution. The requirements for counter electrodes include long-term stability, high conductivity and photocatalytic performance [30]. Some common counter-electrode materials used are platinum-based [31–33], carbon-based [34, 35], polymer composites [36, 37], and transition metal compounds [38].

7.2.1.4 Electrolyte

The electrolyte acts as the charge carrier between the photoanode and counter-electrode. The desirable properties of a DSSC electrode include high charge transport rate, long-term stability, and acceptable dielectric constant. Some common counter-electrode materials can be classified into liquid, solid, and quasi-solid states. Examples of liquid-state electrolytes are organic solvents [39], redox couples [40], and various additives [41, 42]. Examples of solid-state electrolytes include hole transport materials (HTMs) [43] and redox couples incorporating HTMs [44]. Recently, there has been a growing interest in quasi-solid state electrolytes, which include thermoplastic/ thermosetting gels [45, 46] and conducting polymers [47].

MOF-derived components are being explored in DSSC components and have been reported to exhibit exceptional properties. For example, cobalt-doped TiO_2 derived from MIL-125 (Ti) is reported to perform better than the pristine TiO_2-based photoanode [48]. The hierarchically porous TiO_2 nanocrystals are achieved by calcination of the parent MIL-125(Ti), which results in high surface area and pore size. Furthermore, the resultant TiO_2 nanocrystals show quick electron transport, good charge collection efficiency and dye adsorption. Doped TiO_2 nanoparticles derived from MOF were reported to exhibit exceptional photocurrent of about 17.27 mA/cm^2 and photoelectric conversion performance twice as good as the pristine TiO_2 [49]. Ni-doped TiO_2 derived from MIL-125 was reported to show remarkable power conversion efficiency (PCE) [50, 51].

7.2.2 MOF-Derived Materials in PSCs

Perovskites are a class of material with the general formula of ABX_3 where A is the cation, B is the metal cation, and X is the halide atom. Perovskites are named after a Russian mineralogist, Lev Perovski, after its discovery in the Ural Mountains in Russia by mineralogist Gustavus Rose. Since then, it has garnered various interests in multiple fields, including solar cells. In fact, perovskite solar cells are among the top

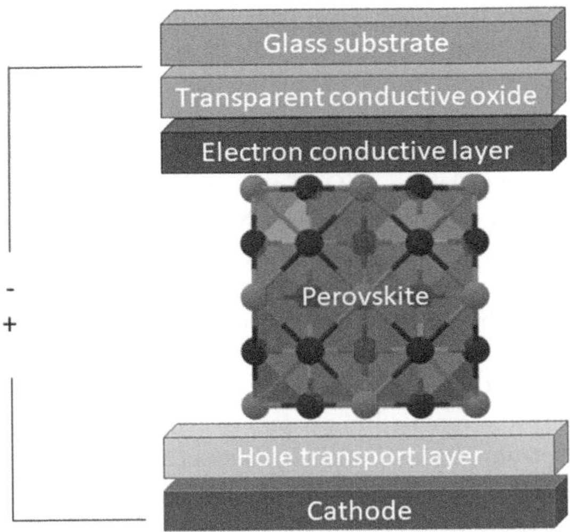

FIGURE 7.3   Working of a perovskite solar cell (PSC).

subsets of solar cells with high reported efficiencies (up to 25.5%) [52]. Perovskites Solar Cells (PSC) contain a conductive substrate, electron transport layer, hole transport layer, light absorber, and a metal electrode. The construction and working of the organic solar cell are as illustrated in Figure 7.3.

The light incident on the PSCs makes the perovskite layer produce excitons (electron-hole pairs). These excitons can further generate current or recombine. The recombination probability in the perovskite materials is reported to be very low, and this feature of longer carrier life span and high diffusion distance is attributed to the high performance of PSCs. The free electrons reach the electron transport material, whereas the free holes reach the hole transport materials. The conductive substrate collects electrons, and the holes are collected at the metal electrode. This leads to photocurrent generation as the system is connected using a back circuit. The interfacial layer is often added between the perovskite and charge transport layers to improve performance [53].

MOFs and MOF-derived materials are extensively studied for their application towards improving perovskite solar cells, especially in the active layers. Some commonly employed materials [54] and their alternative MOF-derived materials (in the case of active layers) are listed below.

1. Conductive substrate – Transparent conducting glass, usually covered with indium tin oxide (ITO) or fluorine-doped tin oxide (FTO), is employed as the conductive substrate.
2. Light absorber – Perovskites, lead-free perovskites, and quantum dots perovskites. This forms the crucial part where perovskite-based materials outshine the rest of the materials.
3. Back electrode/counter electrode – Cu, Au, Ag, Al, carbon, transition metals (Ni, Co, Mo, W, etc.)

Active layers in perovskite solar cells include the electron transport layer, the hole transport layer, and the interfacial layer (if used). MOF-derived materials have many applications in improving the active layers of perovskite solar cells. The materials used in active layers of perovskite materials and their MOF-derived alternatives found in the literature are listed below.

4. Electron transport layer – TiO_2, SnO_2, ZnO, organic ETMs (fullerene based, polymers, small molecules such as NDI and PDI) are common choices for electron transport layers. Electron transport layers are essential as they perform the role of extracting the electrons and avoiding hole transport. The ideal material should have high selectivity towards electrons, mobility, and stability. Gauda et al. recently reported highly efficient solar cells (19.6% power efficiency) using MOF-derived carbon-doped ZnO [55]. The TiO_2, when derived from MOF, were reported to possess excellent properties (for example, Co-doped TiO_2 from NH2-MIL-125 by Bark et al. [56]).

5. Hole transport layer – Commonly employed materials include doped HTMs (spiro-OMeTAD with p-type dopants), dopant-free HTMs (Spiro-OMeTAD with fluorene unit or various ring units), polymeric HTMs (such as PTAA, PTTI, P3HT, PBDT-based), inorganic hole transport materials (including NI-based, Cu-based, and other oxides). Ni-O derived from Ni-based MOFs is employed as a hole transport layer for additional performance improvements. When derived from MOFs, N-rich porous carbon has been demonstrated to be helpful as an auxiliary hole transport layer.

6. Interfacial layer – Commonly employed materials include organic self-assembled monolayers, inorganic oxides, organic polymers and halide perovskites. Highly porous MOF-derived materials have been reported to be efficient interfacial layers. NiO@C nanoparticles derived using MOFs are reported to be improving electron selectivity behaviour. In addition to increased efficiency, reports mention increasing stability by introducing MOF-based materials as interfacial layers.

7.2.3 MOF-Derived Materials in OSCs

Organic solar cells (OSCs) use organic photoactive layers to generate electrical energy from solar energy. Compared to the silicon-based counterpart, the organic active layers are inexpensive, more eco-friendly, and easily manufactured. Nevertheless, the OSCs face the issue of low efficiency. Typically, one or several active layers exist between the two electrodes, which do the charge generation, separation and transport to the back circuit. A bilayer OSC with photoactive layers consisting of a donor and acceptor semiconductor organic materials is a common type of OSV. The OSCs are the ones where the design principles are quite like those of the conventional PN junction cells except for multilayers of organic semiconductors. The incident light makes the donor organic semiconducting material donate electrons, thereby transporting holes. On the other hand, the acceptor organic semiconducting layer withdraws electrons, thereby transporting electrons. The excitons created (electron-hole pair) diffuse to the interface, and a photovoltaic current is generated when the holes and electrons move to the corresponding electrodes.

Although significant research exists exploring MOF-derived materials for both DSSCs and perovskite solar cells, as described in the above sections, there has not been any incorporation in organic solar cells yet. But given the recent interest in pristine MOFs into organic solar cells [57], it shouldn't be too long to see MOF-derived materials with improved performance.

7.3 RECHARGEABLE BATTERIES

MOFs and MOF-derived materials are increasingly researched for their application in next-generation rechargeable batteries such as Lithium-ion batteries (LIBs), Lithium-sulphur batteries (LSBs), Sodium-Ion batteries (SIBs), Potassium-Ion batteries (PIBS), Zinc-Ion batteries (ZIBs), Metal-Air batteries such as Lithium oxygen batteries (LOBs), and Zinc-air batteries.

7.3.1 MOF-Derived Materials for Lithium-Ion Batteries (LIBs)

Lithium-ion batteries are widely popular due to their high-energy battery application towards electric vehicles, intelligent electronics and power grids. The working of a generic LIB is illustrated in Figure 7.4.

Following commercialization by the Sony corporation in 1991 [58], there has been immense research in this field. However, traditional LIBs suffer from severe capacity loss when operated at high charging or discharging rates. Various nanostructured materials are explored to act as electrodes for high-energy LIBs, and MOF-derived materials are no exception in this search.

MOF-derived carbon materials for the electrode form most cases where MOF-derived materials are used in LIBs. MIL-88-Fe has been used as a template to create spindle-like Fe_2O_3 for mesoporous electrodes for LIBs by Cho and co-workers [59]. The MOF-derived carbon material is a promising candidate as it provides an excellent platform to uniformly disperse metal oxides for highly efficient porous LIB electrodes. Qian and co-workers calcined MOF-74 (Ni) to further create NiS@C MOF-derived cathode. Graphite nanocages and other forms of carbon derived from MOFs have been demonstrated to be suitable cathode materials. In terms of the anode,

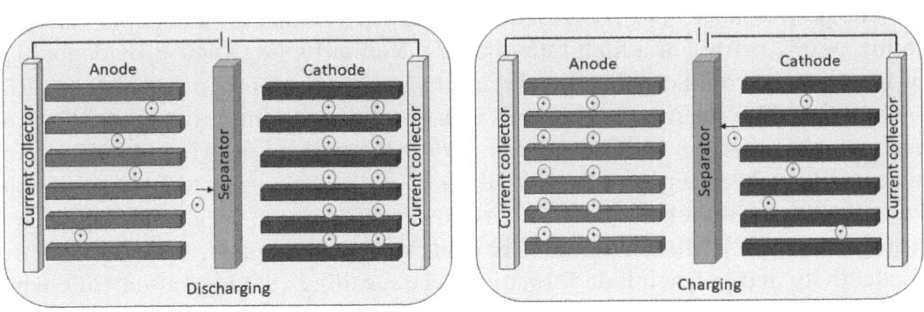

FIGURE 7.4 Charge and discharge operation of a lithium-ion battery (LIB).

MOF-derived composites such as MnO nanocrystals [60], ZnO@C composites [61], CuS@Cu-BTC [62], transition metal oxide composites [63] and CuO@graphene [64] are some of the widely studied systems.

7.3.2 MOF-Derived Materials for Lithium-Sulphur Batteries (LSBs)

Sulphur is an environment-friendly cathode, and lithium-sulphur batteries have gained tremendous interest due to their high theoretical specific capacity and energy density [65]. However, the challenges LSBs face during their application are the low conductivity of sulphur and polysulphide-based intermediate formation during electrochemical reactions. The low conductivity of sulphur leads to low utilization of the sulphur cathode, whereas the creation of polysulphides leads to the deposition of decremental resistive layers. Several methods are being investigated to overcome these challenges [66, 67], one of which is the incorporation of sulphur into porous compounds instead of pure sulphur cathode, which is of high interest. Various MOF-derived materials have been investigated for this purpose.

MOF-derived materials are being explored for cathode and other parts of LSBs. For example, Hao et al. created a MOF-derived microporous carbon polyhedron for improved cathode performance [68]. Specifically, ZIF-8 is heated up to 1000°C, and on annealing, a uniformly distributed morphology is preserved in the creation of microporous carbon polyhedrons (MPCPs). The sulphur loading into micropores of MPCPs was shown to provide high efficiency in electrolytes. It has to be noted that the researchers pointed out that the effect of sulphur loading is significant. While the uniformly dispersed sulphur inside the pores gave advantageous results regarding cathode efficiency, the system demonstrated inferior performance when the composite contained additional sulphur outside the micropores. A similar study on ZIF-8 found that nitrogen-doped carbon (NDC) spheres formed by calcination of the parent MOFS showed excellent electrochemical properties as a cathode for LSBs [69]. One of the problems of LSBs is their poor cycle stability and the use of sulphur itself, as it's not so environmentally friendly after all. In view of this, researchers have used strategies to include polar materials alongside sulphur. This serves the purpose of inhibiting the polysulphide transformation and aids in rapid charge transfers. N doping in the carbon spheres was found to aid in fast charge transfer and sulphur immobilization within the micropores, resulting in an excellent cathode performance. The NDS cathode is reported to display a superior rate capability of 632 mAh/g at a high rate of 5 A/g. Similarly, Fe_3O_4@C provided a high surface area to confine polysulphide and to improve cathode properties in LSBs [70]. A Co-MOF nanosheet was used as a precursor to synthesize novel $Co_{0.85}Se$ nanoparticles that could host sulphur for LSB applications [71]. MOF-derived materials have been reported to be used for LSB electrolytes as well. For example, Poramane and co-workers used UiO as a base structure to create a solid electrolyte for LSBs [72]. Modifications were made to UiO and ligand, allowing selective conductivity and polysulphide formation. The resulting -SO_3Li grafted UiO, when integrated with a Li-based ionic liquid, showed good capabilities while addressing significant issues in LSBs.

7.3.3 MOF-DERIVED MATERIALS FOR SODIUM-ION BATTERIES (SIBs)

Lithium-ion batteries are considered expensive to use for large-scale deployments owing to the limited availability of lithium. On the other hand, sodium is abundant and cheap. Sodium-ion batteries (SIBs) have cost advantages and good electrochemical potential for large-scale deployments. SIBs are competitive to LIBs in many areas and are considered a viable supplement to LIBs for many applications. However, one prominent issue of SIBs is their efficient working temperature (~300°C) [73]. Various improvements are carried out to SIB components, including its anode, electrode and electrolyte. MOF-derived materials have been extensively used in the search for improving SIBs.

MOF-derived materials are being increasingly explored for SIB components. MOF-derived compounds such as heteroatom doped Co [74], N-doped C [75], Fe_7S_8 [76], Na_2FePO_4F-C [77], and $Co(Ni)Se_2$-C [78] for reported for SIB anode. Various other related MOF-derived materials are increasingly being explored to improve SIB properties such as durability, charge transfer rate and storage performance.

7.3.4 MOF-DERIVED MATERIALS FOR POTASSIUM-ION BATTERIES (PIBs)

Potassium-ion batteries show promising avenues for higher-density batteries due to their more negative standard electrode potential and higher operating voltage. In addition to this, the current collector can be replaced with aluminium instead of widely used copper. This is because potassium does not alloy with aluminium and thus reduces cost compared to copper collectors in most other batteries. However, PIBs withhold severe disadvantages of structural instability, rate performances and decomposition reactions. These can be attributed to its high redox potential. MOF-derived materials are researched to improve PIB components, including cathode, electrode and electrolyte.

As with other rechargeable batteries, MOF-derived materials are being reported for components of PIBs. MOF-derived compounds such as $CoSe_2@N$-C [79], carbon nanorods in red phosphorus [80], sulphur-doped flower-like C [81], and N-doped CNTs [82] are some recent advances in this regard. At the current stage, incorporation of MOF-derived materials into PIBs seems to be at a lower rate than in the above battery types.

7.3.5 MOF-DERIVED MATERIALS FOR ZINC-ION BATTERIES (ZIBs)

LIBs have the issue of lithium scarcity around the world. The surge of interest can be seen in low-cost alternatives to LIBs, including SIBs and PIBs. However, Na and K have high activity, which poses safety concerns, and a larger radius poses design concerns [83]. Therefore, a significant rise in interest can be seen in studies related to multivalent earth-abundant cations such as Mg^{2+}, Ca^{2+}, Zn^{2+}, and Al^{3+} due to their increased energy density and better safety-related metrics. However, these polyvalent-based batteries face an issue of electrolyte corrosion of metallic anode and current collectors. In comparison, znc-ion-based batteries (ZIBs) are widely researched due to their improved safety, lower costs, ease of electrolyte selection, higher redox potential, etc. However, ZIBs require significant research for design and upscaling for high-density batteries.

Among the ongoing material research to improve the cathode, anode and electrolyte – MOF-derived materials form a notable class of candidates. Some examples of MOF-derived compounds used in ZIBS include – N-doped C (which is then coated onto MnOx) [84], amorphous V_2O_5 [85], MnO/C hybrid [86], stacked Mn_2O_3@C flakes [87], etc. It should be noted that the ongoing research in this field of MOF-derived materials in ZIBs is pretty compelling and extensive.

7.3.6 MOF-Derived Materials for Metal-Air Batteries

Metal-air batteries are gaining attention due to their higher specific energy compared with the established LIBs [88]. The metal-air batteries consist of a metal anode (called 'anode') and an oxygen-air electrode (called 'cathode'). Metal-air batteries found in literature include usage of alkali metals (such as Li, Na, K) [89, 90], alkaline earth metals (such as Mg) [91], transition metals (such as Fe, Zn) [92, 93], metals (such as Al) [94], mixed-transition metals (such as stannates, ferrates, cobaltates, and nickelates) [95], and metalloids (such as Ge) to some extent [96].

MOF-derived material has been studied for novel anode material, including composites and metal-incorporated porous structures. MOF-derived Co has been reported as a bifunctional catalyst that improves the efficiency of zinc-air batteries [97]. Other example includes MOF-derived Co-N/C nanowires [98], ordered porous C for alkaline-air batteries [99], Ni-Fe doped C [100], Co-N doped C [101], etc. In the current state, metal-air batteries, in general, are being sidelined by other prominent rechargeable batteries seen above. Nevertheless, they are being increasingly improved for efficiency, and clearly, MOF-derived materials are being explored well.

7.4 SUPERCAPACITORS

As the name suggests, supercapacitors (also called electrochemical capacitors) have very high capacitance values. In the regime of energy storage devices, supercapacitors have special interests owing to their high power density, lower internal resistance, lower weight, long life cycle, and, importantly, very high charge and discharge rates [102]. Supercapacitors can be considered as something in between electrolytic capacitors and rechargeable batteries. Supercapacitors have very superior capacitance compared to electrolytic capacitors. On the other hand, they have high power density and higher charging/discharging rates compared to rechargeable batteries. Even though, say, Lithium-ion batteries have ten times more energy density (i.e., they can store more energy per unit mass), supercapacitors have higher power density (i.e., they can store and release more short-term power). Supercapacitors are the best when the application demands a large amount of power more rapidly rather than long-term storage of large amounts of energy, as in the case of batteries. Supercapacitors are used in regenerative braking, burst mode power delivery (such as in military applications), and as a power backup unit [103] (Figure 7.5).

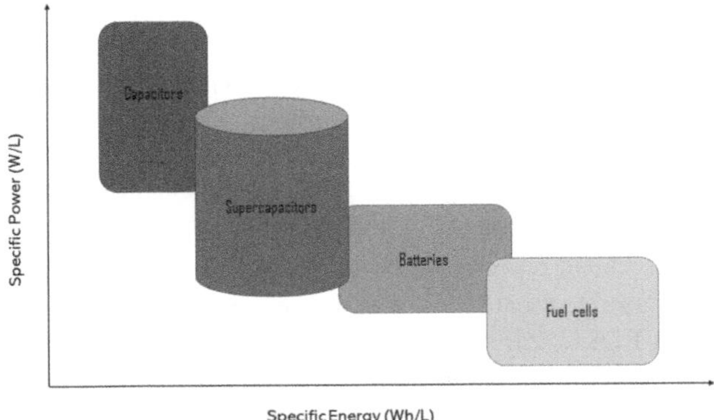

FIGURE 7.5 Supercapacitors compared to capacitors, batteries, and fuel cells.

Supercapacitors can be classified into two categories as follows.

1. Electrical double-layer capacitors (EDLCs): The energy storage in EDLC is done with the help of the electrochemical double layer through a non-faradaic mechanism. Two carbon-based electrodes with high specific surface areas per volume, when immersed in an electrolyte, lead to electrostatic accumulation of charges. The layers of charges accumulated are termed as the "Helmholtz layers".
2. Pseudo-capacitors: The energy storage in pseudo-capacitors is due to the electrochemical faradaic redox reaction between electrolyte ions and electroactive materials. The terminology of pseudo-capacitance comes from the pseudo-capacity used by David C. Grahame in 1941 to describe the excess capacity unrelated to the electrical double layer [104]. Three types of reversible pseudo-capacitance are underpotential deposition, faradaic redox pseudo-capacitance, and intercalation pseudo-capacitance.

MOF-derived materials have been extensively studied to improve the electrode material in both – EDLC and pseudo-capacitors type supercapacitors.

7.4.1 MOF-Derived Electrodes for EDLCs

EDLCs (electrical double-layer capacitors) use the electrostatic effects between two carbon-based electrodes. These carbon-based electrodes, such as nanoporous carbon materials, usually have a high specific surface area per volume and very high charge and discharge rates. Due to their inherent structural advantage, MOFs can leverage the material to have tailorable pore textures and homogeneous nanopore architecture. Therefore, various MOFs have been used as a sacrificial template to create tailored carbon-based nanoporous electrodes to improve supercapacitor performance in the case of EDLC supercapacitors.

Liu et al. reported a MOF-5-derived nanoporous carbon (NPC) from the heat treatment process followed by carbonization into a PFA/MOF-5 composite [105]. Hu et al. reported a high capacitance superconductor when MOF-5, along with a mixture of carbon tetrachloride and ethylenediamine, is heated with KOH [106]. Other than the cases where the MOFs are used as sacrificial templates, there are multiple reports where MOFs themselves are used as the EDLC electrode [107–109]. UiO-66 [110], MOF-74 [111], MOF-808 [112], HKUST-1 [113], and nMOF-867 [114]-based electrodes have also been reported to be employed in EDLC supercapacitors. MIL-101-BH$_2$ has been used to create hollow nanocapsules, like purpose carbon materials, to be employed in EDLC. ZIFs (Zeolitic Imidazolate Framework) are a subclass of MOFs and have been extensively used in EDLC electrodes. This includes ZIFs, ZIF-derived nanoporous carbons, and ZIF-derived composites. ZIF-8 has been used to create nanoporous carbon, whereas ZIF-67 has been reported for deriving nanoporous and Co_3O_4 [115, 116]. In some cases, ZIF-8 and ZIF-67-derived nanoporous carbon materials were reported to be hybridized with polymer to improve morphology and electrical performance. In addition, core-shell ZIF-8@ZIF-67 has reportedly been used to gain the advantageous properties of both the parent ZIF structures [117].

7.4.2 MOF-Derived Electrodes for Pseudo-Capacitors

Pseudo-capacitors rely on faradaic redox reactions between electrolytic ions and the electrolyte. Redox-active metal oxides (MOs) and conducting polymers are common choices for the electrode material in the case of pseudo-capacitors. MOF-derived materials are heavily studied in this segment to derive efficient MOs from being used as pseudo-capacitor electrodes.

Ni-based MOFs are widely adopted to produce pristine MOs. Sun et al. reported nickel oxide nanosheets derived from Ni-based MOFs to have high specific capacity and retain high current density [118]. Han et al. used Ni-MOF to get 2D nickel oxide nanosheets [119]. Li et al. reported MOF precursor to derive a Co_3O_4 micro flower type electrode. Binary metal oxides with two metal cations were studied due to better electrical conductivity and a higher proportion of redox-active sites. Bimetallic MOFs, namely Ni-Co-MOF, have been used to create derived nanosheets for pseudo-capacitor electrodes [120]. Chen and co-workers reported using Zn-Co-MOF to prepare mixed MO as supercapacitors [121]. Some more reports elucidate more usage of MOFs derived mixed metal oxides for pseudo-capacitor electrodes. The mixed metal oxides such as $Ni_xCo_{3-x}O_{4-1}$ (derived from MOF74-Co), Co_3O_4 (derived from ZIF-67), $MnCo_2O_4$ nanocage (derived from Mn-Co-ZIFs), Mn_2NiO_4 (derived from Ni-Mn-MOFs) being studied for pseudo-capacitors electrodes [122].

7.5 HYDROGEN STORAGE

Hydrogen is increasingly being explored for various green energy applications, including hydrogen fuel cells. The main advantage of using hydrogen is its abundance and its superior energy density. On the downside, hydrogen is a very light gas and poses problems in storage and transportation. Hydrogen generation was first demonstrated by scientists William Nicholson and Sir Anthony Carlisle through

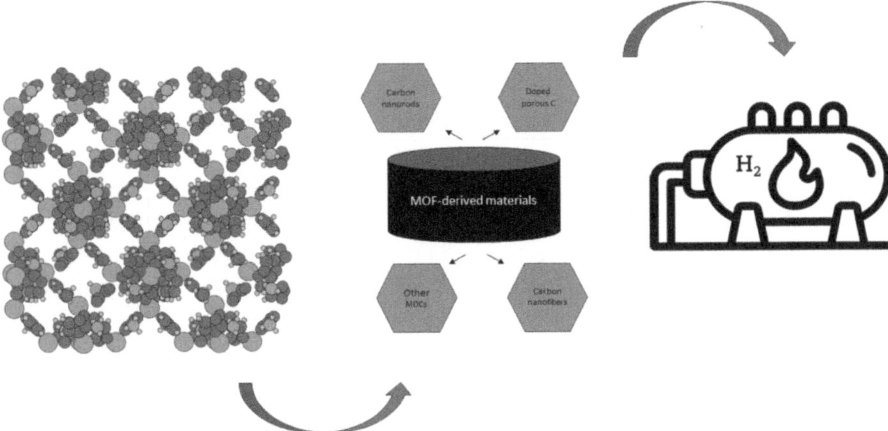

FIGURE 7.6   MOF-derived materials for Hydrogen storage.

electrolysis. One significant milestone in the hydrogen fuel timeline is the dem-
onstration by General Electric for it to be used for the Apollo and Gemini space
missions. Conventional hydrogen storage mainly comprises physical storage systems
including cylinders where the hydrogen is stored as a compressed cryo-compressed
or liquified gas. One material-based approach to storing hydrogen includes porous
materials, and MOF-based porous materials have good capabilities in this regard
(Figure 7.6).

Carbon materials are significant contributors to hydrogen storage materials. Other
than conventional carbon materials such as CNTs, nanocages, nano horns, nano fibres
and graphene, MOF-derived carbon materials have often been reported to exhibit high
porosity and, in many cases, unseen hydrogen uptakes. The hierarchical porosity,
unlike the narrow micropore distribution in regular carbon materials, coupled with
exceptional pore volume, makes MOF-derived carbon materials a strong choice for
hydrogen storage applications. Seung et al. prepared a series of MDCs (MOF-derived
carbon) for reversible hydrogen storage capacities [123]. IRMOF-1, IRMOF-3 and
IRMOF-8 were subjected to a series of pyrolysis steps, including a high-temperature
treatment in a nitrogen environment and cooling to synthesize high crystalline and
porous MDCs. The hydrogen storage capacity of the MDCs was reported to be 30%
higher than their parent MOFs. Therefore, MOF-derived porous carbon can be seen
as an efficient material with exceptional porosity and with the characteristics of
MOFs that deliver better hydrogen storage capabilities. At the same time, this might
not be the case with all MOFs. For example, in a study by Tshiamo et al., nanoporous
carbon composites were derived from both MOF-5 and a chromium-based MOF (Cr-
MOF) [124]. While the porous carbon from the zinc-based MOF-5 showed attractive
hydrogen storage results, the porous carbon from the chromium-based MOF showed
otherwise. The result of this performance was attributed to the lower surface area
and pore volume, which were, in turn, the result of the formation of chromium oxide
and carbide species within the final pores. Table 7.1 lists some central MOF-derived
carbons utilized for hydrogen storage techniques.

TABLE 7.1
MOF-derived carbon materials for hydrogen storage

MOF-derived carbons	Precursor	Adsorbent	Ref.
MDC-1	IRMOF-1	H_2	[123]
MDC-3	IRMOF-3	H_2	[123]
NPC	MOF-5	H_2	[124]
C800	ZIF-8	H_2	[125]
C1000	ZIF-8	H_2	[125]
pCNF	ZIF-8	MgH_2	[126]
NPC	ZIF-67	MgH_2	[127]
NPC	Ni-MOF	MgH_2	[128]

TABLE 7.2
Technical performance requirement for hydrogen storage systems [129]

Technical parameter	2020	2025
Operating ambient temperature	−40°C to 60°C	−40°C to 60°C
Max delivery temperature	−40°C to 85°C	−40°C to 85°C
Min-max delivery pressure	5–12 bar	5–12 bar
Specific energy	1.5 KWh/kg	1.8 KWh/kg
Storage system cost	10 $/kWh	9 $/kWh
Operational cycle life	1500	1500

Magnesium hydride (MgH_2) is another promising candidate for hydrogen storage. MgH_2 confinement studies were made in MOF-derived N-doped carbon nanofibers [126]. Te-ZIF-8 was pyrolyzed to create pCNF (porous carbon nanofibers) under an inert atmosphere. The pCNF were reported to show efficient hydrogen storage even at lower temperatures. Moreover, the cycling stability was reported to be high as the capacity retained was over 95% after ten complete dehydriding and rehydrating cycles. Similarly, MgH_2 storage was strategically improved when using porous composite resulting from ZIF-67 pyrolysis [127]. A carbon-supported transition metal compound, namely $FeCoS_@C$, derived from ZIF-67, acts as a catalytic system for MgH_2 storage. The availability of multiple active sites makes the system undergo enhanced MgH_2 storage.

There is somewhat less research on MOF-derived materials for hydrogen storage solutions. Considering the ongoing need for hydrogen fuel cells and the limits of pressurized cylinder tanks makes studying porous material-based storage systems exciting. Table 7.2 list some major technical performance requirements set by the U.S. Department of Energy along with automotive, energy and utility companies. On the one hand, none of the above-mentioned MDCs or MOF-derived materials has yet achieved the required criteria set by the U.S. Department of Energy for the year 2020. On the other hand, the requirement set for the upcoming 2025 demands materials with ultra-micropores and superior porous capabilities, in which MOF-derived materials have great potential.

7.6 CONCLUSION

MOF-derived materials can be increasingly seen in many energy-related applications focusing on renewables. Active layers of various solar cells, including dye-sensitized solar, perovskite solar cells, and organic solar cells, are found to be replaced with many MOF-derived components. In the case of rechargeable batteries, MOF-derived materials are being explored for alternative cathode, anode, and electrolyte materials due to their hierarchical architecture and increased capabilities. While plenty of research focus is observed in lithium-ion, lithium sulphur, sodium ion, and zinc ion type batteries, potassium ion and metal-air batteries are sidelined. MOF-derived materials can be seen in plenty of supercapacitor components, increasing the efficiency of the electrodes and other related performance metrics. MOF-derived materials are utilized in the case of hydrogen storage technologies as well. There is a high need for more efficient materials to meet the required standards set for hydrogen storage, and MOF-derived materials pose as viable candidates.

REFERENCES

1. Liu, A. Y.; Yan, D.; Phang, S. P.; Cuevas, A.; Macdonald, D. Effective Impurity Gettering by Phosphorus- and Boron-Diffused Polysilicon Passivating Contacts for Silicon Solar Cells. *Sol Energy Mater Sol Cells*, 2018, *179*. https://doi.org/10.1016/j.solmat.2017.11.004.
2. Cui, T.; Lv, R.; Huang, Z. H.; Chen, S.; Zhang, Z.; Gan, X.; Jia, Y.; Li, X.; Wang, K.; Wu, D.; et al. Enhanced Efficiency of Graphene/Silicon Heterojunction Solar Cells by Molecular Doping. *J Mater Chem A Mater*, 2013, *1* (18). https://doi.org/10.1039/c3ta01634j.
3. Miao, X.; Tongay, S.; Petterson, M. K.; Berke, K.; Rinzler, A. G.; Appleton, B. R.; Hebard, A. F. High Efficiency Graphene Solar Cells by Chemical Doping. *Nano Lett*, 2012, *12* (6). https://doi.org/10.1021/nl204414u.
4. Chander, A. H.; Krishna, M.; Srikanth, Y. Comparison of Different Types of Solar Cells – A Review. *IOSR J Electr Electron Eng Ver. I*, 2015, *10*.
5. Jacoby, M. *The Future of Low-Cost Solar Cells*; Chemical & Engineering News, 2016; Vol. 94.
6. Bella, F.; Bongiovanni, R.; Kumar, R. S.; Kulandainathan, M. A.; Stephan, A. M. Light Cured Networks Containing Metal Organic Frameworks as Efficient and Durable Polymer Electrolytes for Dye-Sensitized Solar Cells. *J Mater Chem A Mater*, 2013, *1* (32). https://doi.org/10.1039/c3ta12135f.
7. O'Regan, B.; Grätzel, M. A Low-Cost, High-Efficiency Solar Cell Based on Dye-Sensitized Colloidal TiO2 Films. *Nature*, 1991, *353* (6346). https://doi.org/10.1038/353737a0.

8. Iftikhar, H.; Sonai, G. G.; Hashmi, S. G.; Nogueira, A. F.; Lund, P. D. Progress on Electrolytes Development in Dye-Sensitized Solar Cells. *Materials*, 2019. https://doi.org/10.3390/ma12121998.

9. Hao, S.; Wu, J.; Huang, Y.; Lin, J. Natural Dyes as Photosensitizers for Dye-Sensitized Solar Cell. *Solar Energy* ; 2006; ss*80* (2), 209–214. https://doi.org/10.1016/j.sole018.2005.05.009.

10. El-Agez, T. M.; Taya, S. A.; Elrefi, K. S.; Abdel-Latif, M. S. Dye-Sensitized Solar Cells Using Some Organic Dyes as Photosensitizers. *Optica Applicata*, 2014, *44* (2). https://doi.org/10.5277/oa140215.

11. Hara, K.; Sugihara, H.; Tachibana, Y.; Islam, A.; Yanagida, M.; Sayama, K.; Arakawa, H.; Fujihashi, G.; Horiguchi, T.; Kinoshita, T. Dye-Sensitized Nanocrystalline TiO2 Solar Cells Based on Ruthenium(II) Phenanthroline Complex Photosensitizers. *Langmuir*, 2001, *17* (19). https://doi.org/10.1021/la010343q.

12. Vougioukalakis, G. C.; Philippopoulos, A. I.; Stergiopoulos, T.; Falaras, P. Contributions to the Development of Ruthenium-Based Sensitizers for Dye-Sensitized Solar Cells. *Coord Chem Rev*, 2011. https://doi.org/10.1016/j.ccr.2010.11.006.

13. Chen, C. Y.; Wu, S. J.; Wu, C. G.; Chen, J. G.; Ho, K. C. A Ruthenium Complex with Superhigh Light-Harvesting Capacity for Dye-Sensitized Solar Cells. *Angew Chem Int Ed*, 2006, *45* (35). https://doi.org/10.1002/anie.200601463.

14. Ho, P. Y.; Wang, Y.; Yiu, S. C.; Yu, W. H.; Ho, C. L.; Huang, S. Starburst Triarylamine Donor-Based Metal-Free Photosensitizers for Photocatalytic Hydrogen Production from Water. *Org Lett*, 2017, *19* (5). https://doi.org/10.1021/acs.orglett.7b00042.

15. Naik, P.; Su, R.; Elmorsy, M. R.; Babu, D. D.; El-Shafei, A.; Adhikari, A. V. Molecular Design and Theoretical Investigation of New Metal-Free Heteroaromatic Dyes with D-π-A Architecture as Photosensitizers for DSSC Application. *J Photochem Photobiol A Chem*, 2017, *345*. https://doi.org/10.1016/j.jphotochem.2017.05.033.

16. Wheeler, K. T. Porphyrin Photosensitization. Advances in Experimental Medicine and Biology, Volume 160 Edited by D. Kessel and T. J. Dougherty. *Med Phys*, 1984, *11* (2). https://doi.org/10.1118/1.595483.

17. Sternberg, E. D.; Dolphin, D.; Brückner, C. Porphyrin-Based Photosensitizers for Use in Photodynamic Therapy. *Tetrahedron*, 1998. https://doi.org/10.1016/S0040-4020(98)00015-5.

18. Calza, P.; Vione, D.; Novelli, A.; Pelizzetti, E.; Minero, C. The Role of Nitrite and Nitrate Ions as Photosensitizers in the Phototransformation of Phenolic Compounds in Seawater. *Sci Total Environ*, 2012, *439*. https://doi.org/10.1016/j.scitotenv.2012.09.009.

19. Nowakowska, M.; Kępczyński, M. Polymeric Photosensitizers 2. Photosensitized Oxidation of Phenol in Aqueous Solution. *J Photochem Photobiol A Chem*, 1998, *116* (3). https://doi.org/10.1016/S1010-6030(98)00305-0.

20. Atilgan, S.; Ekmekci, Z.; Dogan, A. L.; Guc, D.; Akkaya, E. U. Water Soluble Distyryl-Boradiazaindacenes as Efficient Photosensitizers for Photodynamic Therapy. *Chem Commun*, 2006, *42*. https://doi.org/10.1039/b612347c.

21. Moczek, Ł.; Nowakowska, M. Novel Water-Soluble Photosensitizers from Chitosan. *Biomacromolecules*, 2007, *8* (2). https://doi.org/10.1021/bm060454+.

22. Bakalova, R.; Ohba, H.; Zhelev, Z.; Ishikawa, M.; Baba, Y. Quantum Dots as Photosensitizers? *Nat Biotechnol*, 2004. https://doi.org/10.1038/nbt1104-1360.

23. Yeoh, M. E.; Chan, K. Y. Recent Advances in Photo-Anode for Dye-Sensitized Solar Cells: A Review. *Int J Energy Res*, 2017. https://doi.org/10.1002/er.3764.

24. Sauvage, F.; Di Fonzo, F.; Li Bassi, A.; Casari, C. S.; Russo, V.; Divitini, G.; Ducati, C.; Bottani, C. E.; Comte, P.; Graetzel, M. Hierarchical TiO2 Photoanode for Dye-Sensitized Solar Cells. *Nano Lett*, 2010, *10* (7). https://doi.org/10.1021/nl101198b.

25. Yu, H.; Zhang, S.; Zhao, H.; Xue, B.; Liu, P.; Will, G. High-Performance TiO2 Photoanode with an Efficient Electron Transport Network for Dye-Sensitized Solar Cells. *J Phys Chem C*, 2009, *113* (36). https://doi.org/10.1021/jp9041974.

26. Li, J.; Zhang, H.; Wang, W.; Qian, Y.; Li, Z. Improved Performance of Dye-Sensitized Solar Cell Based on TiO2 Photoanode with FTO Glass and Film Both Treated by TiCl4. *Physica B Condens Matter*, 2016, *500*. https://doi.org/10.1016/j.physb.2016.07.021.

27. Dhamodharan, P.; Manoharan, C.; Bououdina, M.; Venkadachalapathy, R.; Ramalingam, S. Al-Doped ZnO Thin Films Grown onto ITO Substrates as Photoanode in Dye Sensitized Solar Cell. *Sol Energy*, 2017, *141*. https://doi.org/10.1016/j.sole ner.2016.11.029.

28. Jiang, C. Y.; Sun, X. W.; Lo, G. Q.; Kwong, D. L.; Wang, J. X. Improved Dye-Sensitized Solar Cells with a ZnO-Nanoflower Photoanode. *Appl Phys Lett*, 2007, *90* (26). https://doi.org/10.1063/1.2751588.

29. Lin, C. Y.; Lai, Y. H.; Chen, H. W.; Chen, J. G.; Kung, C. W.; Vittal, R.; Ho, K. C. Highly Efficient Dye-Sensitized Solar Cell with a ZnO Nanosheet-Based Photoanode. *Energy Environ Sci*, 2011, *4* (9). https://doi.org/10.1039/c0ee00587h.

30. Theerthagiri, J.; Senthil, A. R.; Madhavan, J.; Maiyalagan, T. Recent Progress in Non-Platinum Counter Electrode Materials for Dye-Sensitized Solar Cells. *ChemElectroChem*, 2015, *2* (7). https://doi.org/10.1002/celc.201402406.

31. Li, P.; Wu, J.; Lin, J.; Huang, M.; Lan, Z.; Li, Q. Improvement of Performance of Dye-Sensitized Solar Cells Based on Electrodeposited-Platinum Counter Electrode. *Electrochim Acta*, 2008, *53* (12). https://doi.org/10.1016/j.electacta.2007.12.073.

32. Yoon, C. H.; Vittal, R.; Lee, J.; Chae, W. S.; Kim, K. J. Enhanced Performance of a Dye-Sensitized Solar Cell with an Electrodeposited-Platinum Counter Electrode. *Electrochim Acta*, 2008, *53* (6). https://doi.org/10.1016/j.electacta.2007.10.074.

33. Tsekouras, G.; Mozer, A. J.; Wallace, G. G. Enhanced Performance of Dye Sensitized Solar Cells Utilizing Platinum Electrodeposit Counter Electrodes. *J Electrochem Soc*, 2008, *155* (7). https://doi.org/10.1149/1.2919107.

34. Kim, H.; Choi, H.; Hwang, S.; Kim, Y.; Jeon, M. Fabrication and Characterization of Carbon-Based Counter Electrodes Prepared by Electrophoretic Deposition for Dye-Sensitized Solar Cells. *Nanoscale Res Lett*, 2012, *7*. https://doi.org/10.1186/ 1556-276X-7-53.

35. Yen, M. Y.; Yen, C. Y.; Liao, S. H.; Hsiao, M. C.; Weng, C. C.; Lin, Y. F.; Ma, C. C. M.; Tsai, M. C.; Su, A.; Ho, K. K.; et al. A Novel Carbon-Based Nanocomposite Plate as a Counter Electrode for Dye-Sensitized Solar Cells. *Compos Sci Technol*, 2009, *69* (13). https://doi.org/10.1016/j.compscitech.2009.06.003.

36. Xia, J.; Chen, L.; Yanagida, S. Application of Polypyrrole as a Counter Electrode for a Dye-Sensitized Solar Cell. *J Mater Chem*, 2011, *21* (12). https://doi.org/10.1039/c0j m04116e.

37. Kwon, J.; Ganapathy, V.; Kim, Y. H.; Song, K. D.; Park, H. G.; Jun, Y.; Yoo, P. J.; Park, J. H. Nanopatterned Conductive Polymer Films as a Pt, TCO-Free Counter Electrode for Low-Cost Dye-Sensitized Solar Cells. *Nanoscale*, 2013, *5* (17). https://doi.org/ 10.1039/c3nr01294h.

38. Iqbal, M. Z.; Nabi, J. U.; Siddique, S.; Awan, H. T. A.; Haider, S. S.; Sulman, M. Role of Graphene and Transition Metal Dichalcogenides as Hole Transport Layer and Counter Electrode in Solar Cells. *Int J Energy Res*, 2020. https://doi.org/10.1002/ er.5040.

39. Kim, J. H.; Kim, D. H.; So, J. H.; Koo, H. J. Toward Eco-Friendly Dye-Sensitized Solar Cells (DSSCs): Natural Dyes and Aqueous Electrolytes. *Energies*, 2022. https:// doi.org/10.3390/en15010219.

40. Daeneke, T.; Uemura, Y.; Duffy, N. W.; Mozer, A. J.; Koumura, N.; Bach, U.; Spiccia, L. Aqueous Dye-Sensitized Solar Cell Electrolytes Based on the Ferricyanide-Ferrocyanide Redox Couple. *Adv Mater*, 2012, *24* (9). https://doi.org/10.1002/adma.201104837.

41. Kusama, H.; Arakawa, H. Influence of Alkylaminopyridine Additives in Electrolytes on Dye-Sensitized Solar Cell Performance. *Sol Energy Mater Sol Cells*, 2004, *81* (1). https://doi.org/10.1016/j.solmat.2003.09.001.

42. Chang, W. C.; Sie, S. Y.; Yu, W. C.; Lin, L. Y.; Yu, Y. J. Preparation of Nano-Composite Gel Electrolytes with Metal Oxide Additives for Dye-Sensitized Solar Cells. *Electrochim Acta*, 2016, *212*. https://doi.org/10.1016/j.electacta.2016.07.009.

43. Peedikakkandy, L.; Naduvath, J.; Mallick, S.; Bhargava, P. Lead Free, Air Stable Perovskite Derivative Cs2SnI6 as HTM in DSSCs Employing TiO2 Nanotubes as Photoanode. *Mater Res Bull*, 2018, *108*. https://doi.org/10.1016/j.materresbull.2018.08.046.

44. Upadhyaya, H. M.; Senthilarasu, S.; Hsu, M. H.; Kumar, D. K. Recent Progress and the Status of Dye-Sensitised Solar Cell (DSSC) Technology with State-of-the-Art Conversion Efficiencies. *Sol Energy Mater Sol Cells*, 2013, *119*. https://doi.org/10.1016/j.solmat.2013.08.031.

45. Wu, J.; Hao, S.; Lan, Z.; Lin, J.; Huang, M.; Huang, Y.; Fang, L.; Yin, S.; Sato, T. A Thermoplastic Gel Electrolyte for Stable Quasi-Solid-State Dye-Sensitized Solar Cells. *Adv Funct Mater*, 2007, *17* (15). https://doi.org/10.1002/adfm.200600621.

46. Wu, J.; Lan, Z.; Lin, J.; Huang, M.; Hao, S.; Sato, T.; Yin, S. A Novel Thermosetting Gel Electrolyte for Stable Quasi-Solid-State Dye-Sensitized Solar Cells. *Adv Mater*, 2007, *19* (22). https://doi.org/10.1002/adma.200602886.

47. Komiya, R.; Han, L.; Yamanaka, R.; Islam, A.; Mitate, T. Highly Efficient Quasi-Solid State Dye-Sensitized Solar Cell with Ion Conducting Polymer Electrolyte. *J Photochem Photobiol A Chem*, 2004, *164* (1–3). https://doi.org/10.1016/j.jphotochem.2003.11.015.

48. Krishnapriya, R.; Nizamudeen, C.; Saini, B.; Mozumder, M. S.; Sharma, R. K.; Mourad, A. H. I. MOF-Derived Co2+-Doped TiO2 Nanoparticles as Photoanodes for Dye-Sensitized Solar Cells. *Sci Rep*, 2021, *11* (1). https://doi.org/10.1038/s41598-021-95844-4.

49. Tang, R.; Xie, Z.; Zhou, S.; Zhang, Y.; Yuan, Z.; Zhang, L.; Yin, L. Cu2ZnSnS4 Nanoparticle Sensitized Metal-Organic Framework Derived Mesoporous TiO2 as Photoanodes for High-Performance Dye-Sensitized Solar Cells. *ACS Appl Mater Interfaces*, 2016, *8* (34). https://doi.org/10.1021/acsami.6b06183.

50. Dou, J.; Li, Y.; Xie, F.; Ding, X.; Wei, M. Metal-Organic Framework Derived Hierarchical Porous Anatase TiO2 as a Photoanode for Dye-Sensitized Solar Cell. *Cryst Growth Des*, 2016, *16* (1). https://doi.org/10.1021/acs.cgd.5b01003.

51. Nizamudeen, C.; Krishnapriya, R.; Mozumder, M. S.; Mourad, A. H. I.; Ramachandran, T. Photovoltaic Performance of MOF-Derived Transition Metal Doped Titania-Based Photoanodes for DSSCs. *Sci Rep*, 2023, *13* (1). https://doi.org/10.1038/s41598-023-33565-6.

52. Min, H.; Lee, D. Y.; Kim, J.; Kim, G.; Lee, K. S.; Kim, J.; Paik, M. J.; Kim, Y. K.; Kim, K. S.; Kim, M. G.; et al. Perovskite Solar Cells with Atomically Coherent Interlayers on SnO2 Electrodes. *Nature*, 2021, *598* (7881). https://doi.org/10.1038/s41586-021-03964-8.

53. Cho, A. N.; Park, N. G. Impact of Interfacial Layers in Perovskite Solar Cells. *ChemSusChem*, 2017. https://doi.org/10.1002/cssc.201701095.

54. Shao, J. Y.; Li, D.; Shi, J.; Ma, C.; Wang, Y.; Liu, X.; Jiang, X.; Hao, M.; Zhang, L.; Liu, C.; et al. Recent Progress in Perovskite Solar Cells: Material Science. *Science China Chemistry*, 2023. https://doi.org/10.1007/s11426-022-1445-2.

55. Abd El-Lateef, H. M.; Khalaf, M. M.; Heakal, F. E.; Abou Taleb, M. F.; Gouda, M. Electron Transport Materials Based on ZnO@carbon Derived Metal-Organic Framework for High-Performance Perovskite Solar Cell. *Sol Energy*, 2023, *253*. https://doi.org/10.1016/j.solener.2023.02.055.

56. Nguyen, T. M. H.; Bark, C. W. Synthesis of Cobalt-Doped TiO2 Based on Metal-Organic Frameworks as an Effective Electron Transport Material in Perovskite Solar Cells. *ACS Omega*, 2020, *5* (5). https://doi.org/10.1021/acsomega.9b03507.

57. Dou, J.; Chen, Q. MOFs in Emerging Solar Cells. *Chin J Chem*, 2023. https://doi.org/10.1002/cjoc.202200651.

58. Julien, C.; Mauger, A.; Vijh, A.; Zaghib, K. *Lithium Batteries: Science and Technology*; Springer; 2015. https://doi.org/10.1007/978-3-319-19108-9.

59. Xu, X.; Cao, R.; Jeong, S.; Cho, J. Spindle-like Mesoporous α-Fe2O3 Anode Material Prepared from MOF Template for High-Rate Lithium Batteries. *Nano Lett*, 2012, *12* (9). https://doi.org/10.1021/nl302618s.

60. Zheng, F.; Xia, G.; Yang, Y.; Chen, Q. MOF-Derived Ultrafine MnO Nanocrystals Embedded in a Porous Carbon Matrix as High-Performance Anodes for Lithium-Ion Batteries. *Nanoscale*, 2015, *7* (21). https://doi.org/10.1039/c5nr00528k.

61. Yue, H.; Shi, Z.; Wang, Q.; Cao, Z.; Dong, H.; Qiao, Y.; Yin, Y.; Yang, S. MOF-Derived Cobalt-Doped ZnO@C Composites as a High-Performance Anode Material for Lithium-Ion Batteries. *ACS Appl Mater Interfaces*, 2014, *6* (19). https://doi.org/10.1021/am5046873.

62. Wang, P.; Shen, M.; Zhou, H.; Meng, C.; Yuan, A. MOF-Derived CuS@Cu-BTC Composites as High-Performance Anodes for Lithium-Ion Batteries. *Small*, 2019, *15* (47). https://doi.org/10.1002/smll.201903522.

63. Tan, X.; Wu, Y.; Lin, X.; Zeb, A.; Xu, X.; Luo, Y.; Liu, J. Application of MOF-Derived Transition Metal Oxides and Composites as Anodes for Lithium-Ion Batteries. *Inorg Chem Front*, 2020. https://doi.org/10.1039/d0qi00929f.

64. Ji, D.; Zhou, H.; Tong, Y.; Wang, J.; Zhu, M.; Chen, T.; Yuan, A. Facile Fabrication of MOF-Derived Octahedral CuO Wrapped 3D Graphene Network as Binder-Free Anode for High Performance Lithium-Ion Batteries. *Chem Eng J*, 2017, *313*. https://doi.org/10.1016/j.cej.2016.11.063.

65. Seh, Z. W.; Li, W.; Cha, J. J.; Zheng, G.; Yang, Y.; McDowell, M. T.; Hsu, P. C.; Cui, Y. Sulphur-TiO2 Yolk-Shell Nanoarchitecture with Internal Void Space for Long-Cycle Lithium-Sulphur Batteries. *Nat Commun*, 2013, *4*. https://doi.org/10.1038/ncomms2327.

66. Deshmukh, A.; Thripuranthaka, M.; Chaturvedi, V.; Das, A. K.; Shelke, V.; Shelke, M. V. A Review on Recent Advancements in Solid State Lithium-Sulfur Batteries: Fundamentals, Challenges, and Perspectives. *Prog Energy*, 2022. https://doi.org/10.1088/2516-1083/ac78bd.

67. Ji, X.; Evers, S.; Black, R.; Nazar, L. F. Stabilizing Lithium-Sulphur Cathodes Using Polysulphide Reservoirs. *Nat Commun*, 2011, *2* (1). https://doi.org/10.1038/ncomms1293.

68. Wu, H. Bin; Wei, S.; Zhang, L.; Xu, R.; Hng, H. H.; Lou, X. W. Embedding Sulfur in MOF-Derived Microporous Carbon Polyhedrons for Lithium-Sulfur Batteries. *Chem Eur J*, 2013, *19* (33). https://doi.org/10.1002/chem.201301689.

69. Li, Z.; Yin, L. Nitrogen-Doped MOF-Derived Micropores Carbon as Immobilizer for Small Sulfur Molecules as a Cathode for Lithium Sulfur Batteries with Excellent

Electrochemical Performance. *ACS Appl Mater Interfaces*, 2015, *7* (7). https://doi.org/10.1021/am507660y.

70. Liu, G.; Feng, K.; Cui, H.; Li, J.; Liu, Y.; Wang, M. MOF Derived In-Situ Carbon-Encapsulated Fe3O4@C to Mediate Polysulfides Redox for Ultrastable Lithium-Sulfur Batteries. *Chem Eng J*, 2020, *381*. https://doi.org/10.1016/j.cej.2019.122652.

71. Xie, Y.; Cao, J.; Wang, X.; Li, W.; Deng, L.; Ma, S.; Zhang, H.; Guan, C.; Huang, W. MOF-Derived Bifunctional Co0.85Se Nanoparticles Embedded in N-Doped Carbon Nanosheet Arrays as Efficient Sulfur Hosts for Lithium-Sulfur Batteries. *Nano Lett*, 2021, *21* (20). https://doi.org/10.1021/acs.nanolett.1c02037.

72. Chiochan, P.; Yu, X.; Sawangphruk, M.; Manthiram, A. A Metal Organic Framework Derived Solid Electrolyte for Lithium–Sulfur Batteries. *Adv Energy Mater*, 2020, *10* (27). https://doi.org/10.1002/aenm.202001285.

73. Hwang, J. Y.; Myung, S. T.; Sun, Y. K. Sodium-Ion Batteries: Present and Future. *Chem Soc Rev*, 2017. https://doi.org/10.1039/c6cs00776g.

74. Kaneti, Y. V.; Zhang, J.; He, Y. B.; Wang, Z.; Tanaka, S.; Hossain, M. S. A.; Pan, Z. Z.; Xiang, B.; Yang, Q. H.; Yamauchi, Y. Fabrication of an MOF-Derived Heteroatom-Doped Co/CoO/Carbon Hybrid with Superior Sodium Storage Performance for Sodium-Ion Batteries. *J Mater Chem A Mater*, 2017, *5* (29). https://doi.org/10.1039/c7ta03939e.

75. Li, W.; Hu, S.; Luo, X.; Li, Z.; Sun, X.; Li, M.; Liu, F.; Yu, Y. Confined Amorphous Red Phosphorus in MOF-Derived N-Doped Microporous Carbon as a Superior Anode for Sodium-Ion Battery. *Adv Mater*, 2017, *29* (16). https://doi.org/10.1002/adma.201605820.

76. Wang, Y.; Wen, Z.; Wang, C. C.; Yang, C. C.; Jiang, Q. MOF-Derived Fe7S8 Nanoparticles/N-Doped Carbon Nanofibers as an Ultra-Stable Anode for Sodium-Ion Batteries. *Small*, 2021, *17* (38). https://doi.org/10.1002/smll.202102349.

77. Li, H.; Wang, T.; Wang, X.; Li, G.; Shen, J.; Chai, J. MOF-Derived Al-Doped Na2FePO4F/Mesoporous Carbon Nanonetwork Composites as High-Performance Cathode Material for Sodium-Ion Batteries. *Electrochim Acta*, 2021, *373*. https://doi.org/10.1016/j.electacta.2021.137905.

78. Xu, Z.; Huang, Y.; Chen, C.; Ding, L.; Zhu, Y.; Zhang, Z.; Guang, Z. MOF-Derived Hollow Co(Ni)Se2/N-Doped Carbon Composite Material for Preparation of Sodium Ion Battery Anode. *Ceram Int*, 2020, *46* (4). https://doi.org/10.1016/j.ceramint.2019.10.181.

79. Yang, S. H.; Park, S. K.; Kang, Y. C. MOF-Derived CoSe2@N-Doped Carbon Matrix Confined in Hollow Mesoporous Carbon Nanospheres as High-Performance Anodes for Potassium-Ion Batteries. *Nanomicro Lett*, 2021, *13* (1). https://doi.org/10.1007/s40820-020-00539-6.

80. Sui, X.; Huang, X.; Pu, H.; Wang, Y.; Chen, J. Tailoring MOF-Derived Porous Carbon Nanorods Confined Red Phosphorous for Superior Potassium-Ion Storage. *Nano Energy*, 2021, *83*. https://doi.org/10.1016/j.nanoen.2021.105797.

81. Zuo, Y.; Li, P.; Zang, R.; Wang, S.; Man, Z.; Li, P.; Wang, S.; Zhou, W. Sulfur-Doped Flowerlike Porous Carbon Derived from Metal-Organic Frameworks as a High-Performance Potassium-Ion Battery Anode. *ACS Appl Energy Mater*, 2021, *4* (3). https://doi.org/10.1021/acsaem.0c02799.

82. Xiong, P.; Zhao, X.; Xu, Y. Nitrogen-Doped Carbon Nanotubes Derived from Metal–Organic Frameworks for Potassium-Ion Battery Anodes. *ChemSusChem*, 2018, *11* (1). https://doi.org/10.1002/cssc.201701759.

83. Ming, J.; Guo, J.; Xia, C.; Wang, W.; Alshareef, H. N. Zinc-Ion Batteries: Materials, Mechanisms, and Applications. *Mater Sci Eng R: Rep.* 2019. https://doi.org/10.1016/j.mser.2018.10.002.

84. Fu, Y.; Wei, Q.; Zhang, G.; Wang, X.; Zhang, J.; Hu, Y.; Wang, D.; Zuin, L.; Zhou, T.; Wu, Y.; et al. High-Performance Reversible Aqueous Zn-Ion Battery Based on Porous MnOx Nanorods Coated by MOF-Derived N-Doped Carbon. *Adv Energy Mater*, 2018, *8* (26). https://doi.org/10.1002/aenm.201801445.

85. Deng, S.; Yuan, Z.; Tie, Z.; Wang, C.; Song, L.; Niu, Z. Electrochemically Induced Metal–Organic-Framework-Derived Amorphous V2O5 for Superior Rate Aqueous Zinc-Ion Batteries. *Angew Chem Int Ed*, 2020, *59* (49). https://doi.org/10.1002/anie.202010287.

86. Sun, K.; Pang, J.; Zheng, Y.; Xing, F.; Jiang, R.; Min, J.; Ye, J.; Wang, L.; Luo, Y.; Gu, T.; et al. Oxygen Vacancies Enriched MOF-Derived MnO/C Hybrids for High-Performance Aqueous Zinc Ion Battery. *J Alloys Compd*, 2022, *923*. https://doi.org/10.1016/j.jallcom.2022.166470.

87. Liu, C.; Li, Q.; Sun, H.; Wang, Z.; Gong, W.; Cong, S.; Yao, Y.; Zhao, Z. MOF-Derived Vertically Stacked Mn2O3@C Flakes for Fiber-Shaped Zinc-Ion Batteries. *J Mater Chem A Mater*, 2020, *8* (45). https://doi.org/10.1039/d0ta09212f.

88. Blurton, K. F.; Sammells, A. F. Metal/Air Batteries: Their Status and Potential – a Review. *J Power Sources*, 1979, *4* (4). https://doi.org/10.1016/0378-7753(79)80001-4.

89. Zhang, X.; Wang, X. G.; Xie, Z.; Zhou, Z. Recent Progress in Rechargeable Alkali Metal–Air Batteries. *Green Energy Environ*, 2016. https://doi.org/10.1016/j.gee.2016.04.004.

90. Xiang, J.; Yang, L.; Yuan, L.; Yuan, K.; Zhang, Y.; Huang, Y.; Lin, J.; Pan, F.; Huang, Y. Alkali-Metal Anodes: From Lab to Market. *Joule*, 2019. https://doi.org/10.1016/j.joule.2019.07.027.

91. Tong, F.; Wei, S.; Chen, X.; Gao, W. Magnesium Alloys as Anodes for Neutral Aqueous Magnesium-Air Batteries. *J Magnesium Alloys*, 2021. https://doi.org/10.1016/j.jma.2021.04.011.

92. Hang, B. T.; Eashira, M.; Watanabe, I.; Okada, S.; Yamaki, J. I.; Yoon, S. H.; Mochida, I. The Effect of Carbon Species on the Properties of Fe/C Composite for Metal-Air Battery Anode. *J Power Sources*, 2005, *143* (1–2). https://doi.org/10.1016/j.jpowsour.2004.11.044.

93. Caramia, V.; Bozzini, B. Materials Science Aspects of Zinc-Air Batteries: A Review. *Mater Renew Sustain Energy*, 2014, *3* (2). https://doi.org/10.1007/s40243-014-0028-3.

94. Liu, Y.; Sun, Q.; Li, W.; Adair, K. R.; Li, J.; Sun, X. A Comprehensive Review on Recent Progress in Aluminum–Air Batteries. *Green Energy Environ*, 2017. https://doi.org/10.1016/j.gee.2017.06.006.

95. Zhao, Y.; Li, X.; Yan, B.; Xiong, D.; Li, D.; Lawes, S.; Sun, X. L. Recent Developments and Understanding of Novel Mixed Transition-Metal Oxides as Anodes in Lithium Ion Batteries. *Adv Energy Mater*, 2016. https://doi.org/10.1002/aenm.201502175.

96. Wu, S.; Han, C.; Iocozzia, J.; Lu, M.; Ge, R.; Xu, R.; Lin, Z. Germanium-Based Nanomaterials for Rechargeable Batteries. *Angew Chem Int Ed*, 2016. https://doi.org/10.1002/anie.201509651.

97. Hilal, M. E.; Aboulouard, A.; Akbar, A. R.; Younus, H. A.; Horzum, N.; Verpoort, F. Progress of Mof-Derived Functional Materials toward Industrialization in Solar Cells and Metal-Air Batteries. *Catalysts*, 2020. https://doi.org/10.3390/catal10080897.

98. Chen, Y. N.; Guo, Y.; Cui, H.; Xie, Z.; Zhang, X.; Wei, J.; Zhou, Z. Bifunctional Electrocatalysts of MOF-Derived Co-N/C on Bamboo-like MnO Nanowires for

High-Performance Liquid- and Solid-State Zn-Air Batteries. *J Mater Chem A Mater*, 2018, *6* (20). https://doi.org/10.1039/c8ta01859f.

99. Sun, Q.; Zhu, K.; Ji, X.; Chen, D.; Han, C.; Li, T. T.; Hu, Y.; Huang, S.; Qian, J. MOF-Derived Three-Dimensional Ordered Porous Carbon Nanomaterial for Efficient Alkaline Zinc-Air Batteries. *Sci China Mater*, 2022, *65* (6). https://doi.org/10.1007/s40843-021-1933-4.

100. Li, Y. yun; Zou, Q.; Li, Z.; Xie, D.; Niu, Y.; Zou, J.; Zeng, X.; Huang, J. MOF Derived Ni-Fe Based Alloy Carbon Materials for Efficient Bifunctional Electrocatalysts Applied in Zn-Air Battery. *Appl Surf Sci*, 2022, *572*. https://doi.org/10.1016/j.apsusc.2021.151286.

101. Li, J. C.; Wu, X. T.; Chen, L. J.; Li, N.; Liu, Z. Q. Bifunctional MOF-Derived Co-N-Doped Carbon Electrocatalysts for High-Performance Zinc-Air Batteries and MFCs. *Energy*, 2018, *156*. https://doi.org/10.1016/j.energy.2018.05.096.

102. Wu, S.; Liu, J.; Wang, H.; Yan, H. A Review of Performance Optimization of MOF-Derived Metal Oxide as Electrode Materials for Supercapacitors. *Int J Energy Res*, 2019. https://doi.org/10.1002/er.4232.

103. Şahin, M. E.; Blaabjerg, F.; Sangwongwanich, A. A Comprehensive Review on Supercapacitor Applications and Developments. *Energies*, 2022. https://doi.org/10.3390/en15030674.

104. Grahame, D. C. Properties of the Electrical Double Layer at a Mercury Surface. I. Methods of Measurement and Interpretation of Results. *J Am Chem Soc*, 1941, *63* (5). https://doi.org/10.1021/ja01850a014.

105. Liu, B.; Shioyama, H.; Akita, T.; Xu, Q. Metal-Organic Framework as a Template for Porous Carbon Synthesis. *J Am Chem Soc*, 2008, *130* (16). https://doi.org/10.1021/ja7106146.

106. Hu, J.; Wang, H.; Gao, Q.; Guo, H. Porous Carbons Prepared by Using Metal-Organic Framework as the Precursor for Supercapacitors. *Carbon N Y*, 2010, *48* (12). https://doi.org/10.1016/j.carbon.2010.06.008.

107. Gao, W.; Chen, D.; Quan, H.; Zou, R.; Wang, W.; Luo, X.; Guo, L. Fabrication of Hierarchical Porous Metal-Organic Framework Electrode for Aqueous Asymmetric Supercapacitor. *ACS Sustain Chem Eng*, 2017, *5* (5). https://doi.org/10.1021/acssuschemeng.7b00112.

108. Zhang, D.; Shi, H.; Zhang, R.; Zhang, Z.; Wang, N.; Li, J.; Yuan, B.; Bai, H.; Zhang, J. Quick Synthesis of Zeolitic Imidazolate Framework Microflowers with Enhanced Supercapacitor and Electrocatalytic Performances. *RSC Adv*, 2015, *5* (72). https://doi.org/10.1039/c5ra08226a.

109. Choi, K. M.; Jeong, H. M.; Park, J. H.; Zhang, Y. B.; Kang, J. K.; Yaghi, O. M. Supercapacitors of Nanocrystalline Metal-Organic Frameworks. *ACS Nano*, 2014, *8* (7). https://doi.org/10.1021/nn5027092.

110. Tan, Y.; Zhang, W.; Gao, Y.; Wu, J.; Tang, B. Facile Synthesis and Supercapacitive Properties of Zr-Metal Organic Frameworks (UiO-66). *RSC Adv*, 2015, *5* (23). https://doi.org/10.1039/c4ra11896k.

111. Pachfule, P.; Shinde, D.; Majumder, M.; Xu, Q. Fabrication of Carbon Nanorods and Graphene Nanoribbons from a Metal-Organic Framework. *Nat Chem*, 2016, *8* (7). https://doi.org/10.1038/nchem.2515.

112. Li, J. H.; Chen, Y. C.; Wang, Y. Sen; Ho, W. H.; Gu, Y. J.; Chuang, C. H.; Song, Y. Da; Kung, C. W. Electrochemical Evolution of Pore-Confined Metallic Molybdenum in a Metal-Organic Framework (MOF) for All-MOF-Based Pseudocapacitors. *ACS Appl Energy Mater*, 2020, *3* (7). https://doi.org/10.1021/acsaem.0c00399.

113. Yan, X.; Li, X.; Yan, Z.; Komarneni, S. Porous Carbons Prepared by Direct Carbonization of MOFs for Supercapacitors. *Appl Surf Sci*, 2014, *308*. https://doi.org/10.1016/j.apsusc.2014.04.160.

114. Xu, B.; Zhang, H.; Mei, H.; Sun, D. Recent Progress in Metal-Organic Framework-Based Supercapacitor Electrode Materials. *Coord Chem Rev*, 2020. https://doi.org/10.1016/j.ccr.2020.213438.

115. Jiang, H. L.; Liu, B.; Lan, Y. Q.; Kuratani, K.; Akita, T.; Shioyama, H.; Zong, F.; Xu, Q. From Metal-Organic Framework to Nanoporous Carbon: Toward a Very High Surface Area and Hydrogen Uptake. *J Am Chem Soc*, 2011, *133* (31). https://doi.org/10.1021/ja203184k.

116. Salunkhe, R. R.; Tang, J.; Kamachi, Y.; Nakato, T.; Kim, J. H.; Yamauchi, Y. Asymmetric Supercapacitors Using 3D Nanoporous Carbon and Cobalt Oxide Electrodes Synthesized from a Single Metal-Organic Framework. *ACS Nano*, 2015, *9* (6). https://doi.org/10.1021/acsnano.5b01790.

117. Tang, J.; Salunkhe, R. R.; Liu, J.; Torad, N. L.; Imura, M.; Furukawa, S.; Yamauchi, Y. Thermal Conversion of Core-Shell Metal-Organic Frameworks: A New Method for Selectively Functionalized Nanoporous Hybrid Carbon. *J Am Chem Soc*, 2015, *137* (4). https://doi.org/10.1021/ja511539a.

118. Sun, Z.; Hui, L.; Ran, W.; Lu, Y.; Jia, D. Facile Synthesis of Two-Dimensional (2D) Nanoporous NiO Nanosheets from Metal-Organic Frameworks with Superior Capacitive Properties. *New J Chem*, 2016, *40* (2). https://doi.org/10.1039/c5nj02261d.

119. Pan, Y.; Han, Y.; Chen, Y.; Li, D.; Tian, Z.; Guo, L.; Wang, Y. Benzoic Acid-Modified 2D Ni-MOF for High-Performance Supercapacitors. *Electrochim Acta*, 2022, *403*. https://doi.org/10.1016/j.electacta.2021.139679.

120. Jiang, C.; Li, S.; Li, B.; Liu, S.; Dong, W. W.; Wu, Y. P.; Tian, Z. F.; Zhang, Q.; Li, D. S. In Situ Synthesis of Hierarchical NiCo-MOF@Ni1−xCox(OH)2heterostructures for Enhanced Pseudocapacitor and Oxygen Evolution Reaction Performances. *Dalton Trans*, 2021, *50* (8). https://doi.org/10.1039/d0dt03872e.

121. Chen, S.; Xue, M.; Li, Y.; Pan, Y.; Zhu, L.; Zhang, D.; Fang, Q.; Qiu, S. Porous ZnCo2O4 Nanoparticles Derived from a New Mixed-Metal Organic Framework for Supercapacitors. *Inorg Chem Front*, 2015, *2* (2). https://doi.org/10.1039/c4qi00167b.

122. Mahmood, A.; Zou, R.; Wang, Q.; Xia, W.; Tabassum, H.; Qiu, B.; Zhao, R. Nanostructured Electrode Materials Derived from Metal-Organic Framework Xerogels for High-Energy-Density Asymmetric Supercapacitor. *ACS Appl Mater Interfaces*, 2016, *8* (3). https://doi.org/10.1021/acsami.5b10725.

123. Yang, S. J.; Kim, T.; Im, J. H.; Kim, Y. S.; Lee, K.; Jung, H.; Park, C. R. MOF-Derived Hierarchically Porous Carbon with Exceptional Porosity and Hydrogen Storage Capacity. *Chem Mater*, 2012, *24* (3). https://doi.org/10.1021/cm202554j.

124. Segakweng, T.; Musyoka, N. M.; Ren, J.; Crouse, P.; Langmi, H. W. Comparison of MOF-5- and Cr-MOF-Derived Carbons for Hydrogen Storage Application. *Res Chem Intermed*, 2016, *42* (5). https://doi.org/10.1007/s11164-015-2338-1.

125. Hirscher, M.; Zhang, L.; Oh, H. Nanoporous Adsorbents for Hydrogen Storage. *Appl Phys A Mater Sci Process*, 2023, *129* (2). https://doi.org/10.1007/s00339-023-06397-4.

126. Ren, L.; Zhu, W.; Zhang, Q.; Lu, C.; Sun, F.; Lin, X.; Zou, J. MgH2 Confinement in MOF-Derived N-Doped Porous Carbon Nanofibers for Enhanced Hydrogen Storage. *Chem Eng J*, 2022, *434*. https://doi.org/10.1016/j.cej.2022.134701.

127. Fu, Y.; Zhang, L.; Li, Y.; Guo, S.; Yu, Z.; Wang, W.; Ren, K.; Peng, Q.; Han, S. Catalytic Effect of MOF-Derived Transition Metal Catalyst FeCoS@C on Hydrogen Storage of Magnesium. *J Mater Sci Technol*, 2023, *138*. https://doi.org/10.1016/j.jmst.2022.08.019.

128. Huang, T.; Huang, X.; Hu, C.; Wang, J.; Liu, H.; Xu, H.; Sun, F.; Ma, Z.; Zou, J.; Ding, W. MOF-Derived Ni Nanoparticles Dispersed on Monolayer MXene as Catalyst for Improved Hydrogen Storage Kinetics of MgH2. *Chem Eng J*, 2021, *421*. https://doi.org/10.1016/j.cej.2020.127851.

129. U.S Department of Energy. DOE Technical Targets for Onboard Hydrogen Storage for Light-Duty Vehicles | Department of Energy. *Energy.gov*, 2017, No. August 2009.

8 Environmental Applications

8.1 INTRODUCTION

Environmental pollution is feared to cause irreversible damage to our planet by 2025 [1]. Environmental pollution can be defined as "the contamination of air, water, or food in such a manner as to cause real or potential harm to human health or well-being, or to damage or harm nonhuman nature without justification" [2]. The effect of environmental pollution on human health is of severe impact, like lung cancer and asthma due to air pollution; cancer and cardiovascular disorders due to water contamination; and bacterial, viral, and parasitic infections due to food contamination. Environmental pollution also causes a significant addition to the economic burden of countries. It also engulfs an invaluable workforce in terms of related research, remediation, and finance. Five significant contributors to environmental contamination are air pollution, water pollution, noise pollution, solid waste pollution, and electromagnetic pollution (Figure 8.1).

MOFs and MOF-derived materials hold a significant space in porous materials employed to remediate environmental pollutants. Their porous architecture, mild synthesis conditions, and flexible and integrative nature make them an attractive candidate for many such pollutant capture applications. MOF-derived materials, in particular, take in some advantages of MOF precursor and translate them to even superior capabilities in several instances. This also includes coupling the benefits from other types of materials when made into composites. MOF-derived materials and their application to environmental remediations will be covered in this chapter. Specifically, MOF-derived materials used for air pollutants, water pollutants, food contamination, noise and electromagnetic pollution will be covered. Finally, greener MOF synthesis practices to alleviate environmental damage will be outlined.

8.2 WATER POLLUTANTS

Climate change and population growth are exacerbating the situation of safe-water scarcity. According to recent reports from WHO, over 2 billion people live in water-stressed countries. Since the dawn of newer technology, governments and activists

FIGURE 8.1 Five major environmental pollution.

have been working hard to make safe water available to most of the public. Over the decades, several countries have reported dramatic reductions in several harmful pollutants such as ammonia, phosphates, and toxic metals. However, this is far from the expected decline in pollutants to make safe drinking water available to the human population. Specifically, we need help eliminating contaminants, such as agricultural fertilizers and industrial chemicals.

Water pollution can be classified differently, including origin, water source affected, treatability, and toxicity. Various organic and inorganic pollutants are affecting the water source and are increasing at an alarming rate. Conventional water pollutants include fluoride, nitrate, phosphorus, and trace heavy metals. On the other hand, a class of emerging water pollutants exists, including pharmaceuticals, artificial food enhancers, surfactants, hormones, and endocrine-disrupting compounds (Figure 8.2).

The water pollutants can be classified as follows for simplicity and relevance to the study.

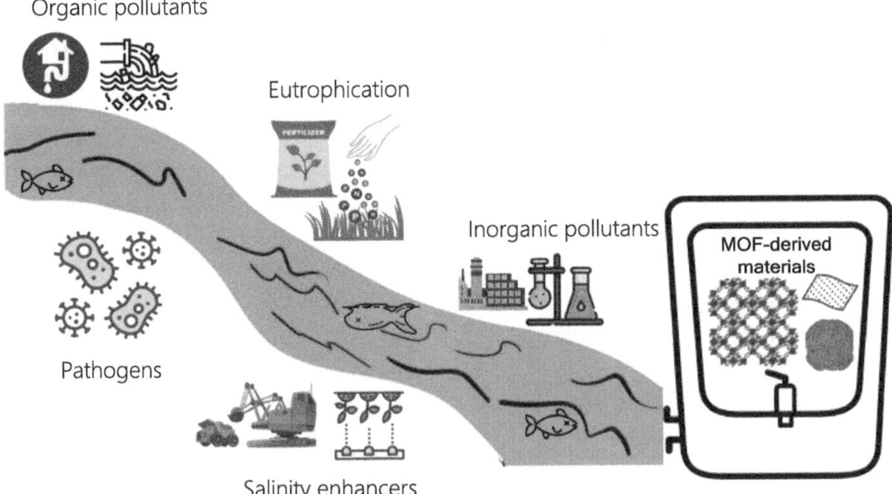

FIGURE 8.2 Major water pollutants.

1. Organic pollutants

   Domestic wastage through sewers and infectious wastage from municipalities are the primary sources of organic pollutants in water sources. These types of decomposable organic wastes are of primary concern due to their deoxygenation nature of the water bodies, causing a significant threat to aquatic life and water quality.

2. Eutrophication

   Agricultural wastages such as fertilizers and pesticides remain one of the top contributors to water pollution. Phosphorous and nitrates are some of the leading contributors in this regard.

3. Other inorganic pollutants

   Other than those from eutrophication, several other inorganic contaminants exist, such as metals and their salts, sulphides, ammonia, oxides of nitrogen, and acids/bases from chemical laboratories.

4. Salinity enhancers

   The salinity of the water source is challenged due to salinity pollution from wastewater originating from irrigation systems, mines and improperly treated water from related factories. One-tenth of all the rivers (in Asia, Africa and America) are reported to be either moderately or severely affected by saline pollution.

5. Pathogen pollution

   Water with high pathogenic content, when used for human consumption or agricultural irrigation, poses a severe threat to health. The primary sources of such pathogenic pollution are the animal industry, human and domestic animal sewage, and wildlife.

Various physical and chemical treatment methods are being studied to advance water purification techniques. These include separation using membrane technologies, adsorption, coagulation, photodegradation, disinfectants, reverse osmosis, UV filtration, etc. Adsorption using porous materials is one of the simple and low-cost solutions for wastewater remediation. While MOFs are widely recognized as effective adsorbents, their practical application for removing contaminants from water bodies is limited by their poor aqueous stability. Therefore, recent years have increased research interest in MOF-derived materials as adsorbents. MOF-derived materials that retain the original MOF's porous structure with additional aqueous stability are promising adsorbent materials. This section outlines various MOF-derived materials employed for water remediation.

8.2.1   MOF-Derived Carbonaceous Materials (MDCs)

Carbonaceous materials include activated carbon, activated carbon fibres and carbon aerogels. Nanoporous carbon materials are found in applications ranging from pollutant capture to energy storage, catalysis and other related areas. They hold a critical stature in the field of environmental remediation. The main advantage of carbon nanoporous materials is their surface area and chemical stability. The major disadvantage is the controllability of their geometry and architecture. When such carbonaceous materials are made from MOFs, they could gain better architecture and textural properties, making them even more attractive for water remediation.

Carbon derived from various MOF materials is reported to show the capabilities of adsorptive removal of emerging contaminants. Five variants of MDCs denoted as CDM-1, CDM-74, CDM-4, CDM-5, and CDM-6, were obtained through carbonization of high valent MOFs (bio-MOF-1(Zn) and MOF-74 (Zn)) and low valent MOFs (metal azolate framework (MAF-4 (Zn), MAF-5 (Zn), and MAF-6 (Zn)) MOF-74 (Zn)), respectively [3]. These MDCs were applied to remove five emerging contaminants, namely, phthalic acid, diethyl phthalate, N,N-diethyl-3-methyl benzamide, chloroxylenol, and oxybenzone from water. Among the different MDCs, CDM-74 exhibited the highest adsorption capacity for removing all the contaminants. Activated carbon produced from the carbonization of Zn-MOF is reported to be effective in removing phosphate ions [4]. Carbonization of Zn-MOF at an optimum carbonization temperature of 500°C is reported to remove phosphate ions from water through monolayer and multilayer adsorptions at higher and lower temperatures, respectively. Though the phosphate removal was slightly higher with Zn-MOF@700°C compared to Zn-MOF@500°C, given the additional energy requirement of carbonization at 700°C, the later sample was reported to be of optimum condition. Carbon derived from Zn-MOF@500°C resulted in 99.47% phosphate removal with an excellent adsorption capacity of 226.07 mg/g compared to 98.4% removal by Zn-MOF with an adsorption capacity of 123.44 mg/g. Doping of carbon derived from MOF is one of the ways to enhance the phosphate ion removal from water further. For example, phosphorous removal is reported to improve when using Fe-N doped carbon derived from MIL-101(Fe) due to a combination of ligand exchange, electrostatic attraction, and electric field effect [5]. A remarkable work by Imteaz Ahmed and co-workers reports the preparation of porous carbon for efficient liquid phase adsorption applications in water and fuel [6]. ZIF-8 was injected with an ionic liquid (IL) via the ship-in-bottle method, which, when pyrolyzed at a higher temperature, gives rise to porous carbon with ionic liquid contents termed IMDCs (ionic liquid@MOF-derived carbon). IMDCs are reported to exhibit remarkable performances when used to adsorb organic contaminants, atrazine (ATZ), diuron, and diclofenac. A well-known organic dye, rhodamine B (RhB), is reported to be efficiently removed by a co-absorbent resulting from pyrolysis of Co-BTC MOF. This particular dye, Rhodamine B (RhB), is widely used in various textile, pharmaceutical and plastic-related industries.

8.2.2 MOF-Derived Magnetic Materials

Much interest can be seen in the research community in developing magnetic materials derived from MOFs for water remediation. The primary reason is that the MOF-derived material can be taken out of action quickly with the help of external magnets. This is quite useful for removing foreign material from the water after adsorption.

The carbonization of MOF-74 (Zn) under a nitrogen atmosphere is reported to yield a porous carbon with excellent porosity and magnetic properties [7]. The porous carbon termed C-MOF-74 (Zn)/carbon rods were found to possess very high dye adsorption capabilities from water sources. The author further went on to modify the magnetic sorbent to improve adsorption characteristics. For example, carbon rods from MOF-74 (Co) were further treated to replace the residual cobalt present in the porous carbon with ZIF-67 by a series of chemical reactions. The resultant ZIF-67@C (derived from MOF-74 (Cu)) was found to exhibit higher extraction abilities of phenolic compounds [8]. Lin et al. prepared a multifunctional magnetic carbon sponge by carbonizing the ZIF-67 precursor. The resulting magnetic sponge is reported to combine the macroporous properties of carbon sponges with the nanoscale porosity and catalytic sites of MOF frameworks. A carbon sponge derived from ZIFs is reported to separate oil from water and emulsions [9]. Similarly, magnetic Co nanoparticles from one-step carbonization of ZIF-67 are reported to retain the shape of the MOF with excellent dye adsorption performance [10]. Jafari et al. synthesized a magnetic N-doped CNT/PC hybrid sorbent material by pyrolysis of loaded ZIF-8 [11]. The magnetic sorbent thus prepared was shown to exhibit good adsorption capacity to separate lead from aqueous water [magnetic 5]. Magnetic porous carbon-based sorbent (MPCS) with core-shell structure was achieved by pyrolysis of modified MOF-5 (Fe). The resultant MPCS was found effective in removing atrazine (ATZ) from water [12]. Hierarchical magnetic carbon composites ($Co_@NC$) derived from the pyrolysis of MOF are reported to be used for separating amodiaquine (ADQ) from water [13]. Catalytic reduction of p-nitrophenol in water is achieved by ferromagnetic cobalt-carbon composites prepared via one-step pyrolysis of ZIF-67 and Co-BTC [14]. In an interesting experimental setup by Lanling Dal and co-workers, ZIF-67 on bamboo fibre bundles (BFB) was pyrolyzed to produce magnetic Co/CoO nanoparticles. These magnetic nanoparticles are shown to effectively remove antibiotics (such as tetracycline hydrochloride) through catalytic degradation mechanisms [15]. For the first time, a Hofmann-type MOF ([CoFe]pyridine[Ni(Cn4)]]) was pyrolyzed to derive CoFeNi alloy nanoparticles with magnetic properties [16]. The CoFeNi alloy nanoparticles are confined within CNTs during experimental setups, resulting in a highly efficient platform for removing a widely used UV stabilizer (BP1 – 2,4-dihydroxybenzophenone (benzophenone-1).

8.3 AIR POLLUTANTS

Air quality has experienced a severe setback throughout the recent decade due to various pollutants. Reports suggest that more than 99% of the human population lives in areas with lower air quality than WHO standards and that around 4 million deaths can be associated with air quality annually. Among the various types of pollutants, the ones with strong evidence for health concern include particulate matter (which includes sulphates, nitrates, ammonia, sodium chloride, black carbon, and mineral dust), nitrogen dioxide, ozone, carbon monoxide, sulphur dioxide, lead, polycyclic aromatic hydrocarbons, formaldehyde, radon, black carbon, ultrafine particles, mould as well as greenhouse emissions (Figure 8.3).

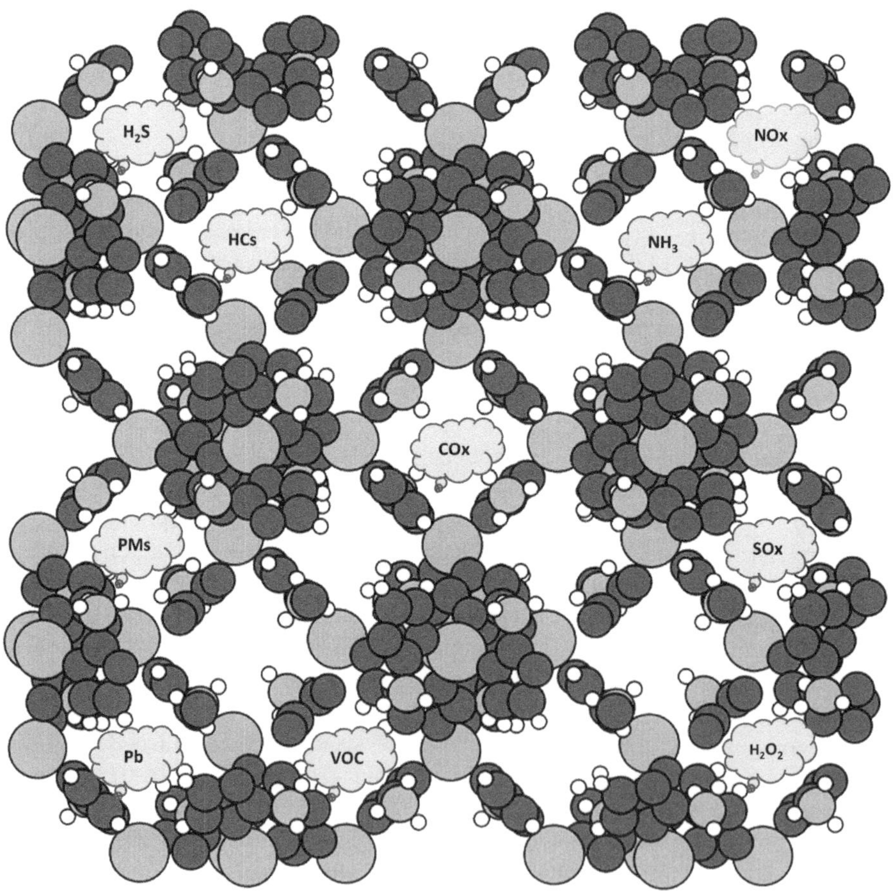

FIGURE 8.3 Major air pollutants.

Some common materials developed for environmental applications include – activated carbon, activated carbon fibres, carbon aerogels, zeolites, MOFs, and MOF-derived materials. The following section focuses on MOF-derived materials used for environmental remediation.

8.3.1 MOF-Derived Activated Carbon

Activated carbon can be described as any porous carbonaceous material with increased surface area and enhanced functional groups. Activated carbon has been historically used as a decolourizing agent, odour eliminator and pollutant capture. Adsorption through the micropores of activated carbon has been a versatile technique employed for various environmental applications. The texture, functional groups, pore size and geometry distributions are some of the physical properties of activated carbon that affect its pollutant adsorption capabilities. Activated carbons are produced from carbonaceous source materials, primarily organic, such as bamboo, coconut husk, peat, petroleum pitch, and lignite. The manufacturing process (termed 'activation') can be physical or chemical. Physical activation is when the carbonaceous source is activated using hot gases. It is achieved using one or more of the following processes.

1. **Carbonization:** The source material is pyrolyzed in the range of 600°C–900°C using an inert gas atmosphere.
2. **Chemical activation:** The source material is carbonized at a temperature of range 600°C–1200°C in an oxidizing atmosphere. Chemical activation is achieved using certain chemicals (mostly strong acids, bases or salt), followed by heating to a high temperature. Based on the preparation methods and applications, the manufactured activated carbon can be classified into one of the following – powdered activated carbon, granular activated carbon, pelletized activated carbon, extruded activated carbon, powdered activated carbon, bead activated carbon, impregnated carbon, polymer coated carbon, catalytic activated carbon and coven carbon.

Activated carbon is reportedly efficient in filtering some volatile organic compounds (isooctane, ethanal, and butyl acetate) from the air [17]. Similarly, activated carbon has been studied for its CO_2 removal abilities [18]. The study also elucidates that the textural properties and the activation method of the activated carbon employed hugely determine the carbon-di-oxide capture ability. The activated carbon injection method has been reported to be used for mercury removal and sulphur dioxide in industrial scenarios [19]. Activated carbon, when made into the form of activated carbon cloth, has been utilized to remove various organic and inorganic gaseous pollutants [20].

MOF-based activated carbons are of increased interest due to the structural advantages of MOF parent structure. Activated carbon, when derived from MOF-based materials, poses additional benefits in structure and properties. Moreover, the availability, cost, ease of application, flexibility and robust surface modification make MOFs a preferred substrate/composite candidate for activated carbon. Cucumber peel-activated carbon from MIL-101 (Cr) is studied for its pollutant adsorption capacities [21]. Dye removal by MOF-5/AC composite was simulated using machine learning by Askari and co-workers [22]. Kayal et al. utilized MIL-101-Cr/AC composite prepared through hydrothermal reaction for methane and carbon dioxide adsorption [23].

8.3.2 MOF-Derived Activated Carbon Fibres

Activated carbon fibres (ACFs) are generally considered a separate class of carbonaceous porous material as they show distinct geometrical properties and applications. For example, the increased voids between the fibres are reported to cause less resistance to the gas that flows through them [24, 25]. ACFs are manufactured from various sources, including natural fibres (cotton, sisal, etc.) and synthetic fibres (such as rayon, resin, and other synthetic fibres). Popular source materials include polyacrylonitrile (PAN), phenolic resin, cellulosic fibres and pitch fibres. ACFs are prepared through a series of steps, such as spinning, stabilizing, carbonizing, and activating precursor fibres. The activation of precursor fibres can, in turn, be a physical or chemical activation procedure. Physical methods are when steam or carbon dioxide is employed as an activation agent. On the other hand, chemical activation is used for cross-linking using silane coupling agents, sulphuric acid treatments and electron radiation.

ACFs have been employed to remove formaldehyde [26, 27]. Activated carbon fibres were used in conjunction with carbon nanotubes to form CNT/ACF composites, which were employed for particulate matter, ozone and other volatile organic compounds [28]. Carbon dioxide removal is studied with nitrogen-containing ACFs at different working temperatures [29]. ACFs, similar to activated carbon, have been used to make various cloths to remove atmospheric pollutants such as toluene, acetaldehyde, methane and carbon dioxide [30, 31]. As in the case of activated carbon, ACFs also demonstrate a high dependence of geometry and other related properties on pollutant adsorption capabilities. Reports suggest various strategies to improve pollutant adsorption, including TiO_2 layer deposition, impregnating with palladium nanoparticles, and decoration with MnO_2 catalyst [32–34].

8.3.3 MOF-Derived Carbon Aerogels

In the area of aerogels, the pioneering work by Pekala and Satcher made the RF (resorcinol-formaldehyde polymer aerogels) aerogels available to the world. Soon after, the researchers at Livermore laboratory discovered carbon aerogel, which was then achieved by heating the RF aerogel and removing the counterparts. RF-based carbon aerogels still form a significant portion of carbon aerogels manufactured for various applications, including electrical conductors, superconductors, energy storage applications, and pollutant capture. Other than RF precursor, many other organic aerogels are reported to be used to produce carbon aerogels using pyrolysis under inert conditions. Some examples of other organic precursors include sol-gel polymerization involving phenol-formaldehyde, melamine-formaldehyde, polyacrylamide, polyurethane, tetraethoxysilane, cellulose esters, and epoxy precursors.

TABLE 8.1
MOF-derived carbonaceous materials for air pollution

Type	Precursor/type	Pollutant adsorption	Ref.
Activated carbon (AC)	Biomass	HO_x, SO_2, H_2S	[39, 40]
	Waste tyres	Methane, SO_2	[41, 42]
	AC with Silver colloidal layers	Microbes, NO, CO, NO_2	[43]
	AC with bio-compost	Biofiltration, BTEX vapours	[44]
	AC with TiO_2	Indoor pollutants	[45]
	AC as a face mask	CO	[46]
	AC and composites	VOCs	[47–50]
	Doped Acs	Carbon dioxide	[51–54]
	MOF-derived Acs	VOCs and others	[21]
Activated carbon fibres (ACFs)	ACFs with DC current	Toluene, Benzene, Limonene, Formaldehyde	[55]
	ACFs with CNT composite	Ozone, particulate matter	[56]
	ACFs cloth	VOCs, HAPs	[57–59]
	Phenolic resin	NO, particulate matter	[60]
	Impregnated ACFs	Acetaldehyde	[61]
	Doped ACFs	Carbon dioxide	[62, 63]
Carbon aerogel (CA)	N-doped CA	Microbes, VOC	[64]
	CA with CNT	Carbon dioxide	[65]
	Newspaper waste	VOCs	[66]
	CA with s-HCA/zein	Bio-facial mask	[67]
	Doped CA	Carbon dioxide	[68–70]

Exceptional carbon dioxide adsorption was reported using carbon aerogel from cellulose [35]. Jianbo Hu et al. reported a carbon aerogel-alkali metal carbonate composite to be capable of high carbon dioxide adsorption from flue gases [36]. Bio-based carbon aerogel prepared using chitosan-polybenzoxazine was reported to pose high carbon dioxide adsorption capacity. Activated carbon aerogels have been reported to be efficient in removing volatile organic compounds, including methanol, acetone, cyclohexane, benzene, toluene, xylenes, acetone, n-Hexane, and sarin [37]. Carbon aerogel-based thin film composite is reportedly employed as a sensing device for volatile vapour at room temperature [38]. Various modifications and composites are being studied to improve the utilization of carbon aerogels for environmental remediation; some of them are outlined in Table 8.1.

8.4 MOF-DERIVED MATERIALS FOR FOOD SAFETY

Food safety is becoming increasingly important with the rise in environmental pollution. As per the world health statistics, 1 in 10 people in the world is prone to sickness from contaminated food, resulting in thousands of deaths annually. Foodborne diseases are often underestimated and are often a catastrophic element in the global economic burden. In general, foodborne illnesses are attributed to bacteria, viruses, parasites, prions and, to our concern, chemicals. In addition to naturally occurring toxins (such as mycotoxins, marine biotoxins, and cyanogenic glycosides), persistent organic pollutants (POPs), heavy metals, radioactive substances and other toxic chemicals are a significant threat to food safety.

The majority of concerns stem from pesticide residues in plants, veterinary drugs in animals, pathogens and mycotoxins, heavy metals, toxic additives, and other contaminants. MOF-derived materials are used in numerous sensing applications to aid food safety in the forms of luminescent sensors, electrochemical sensors, electrochemiluminescence (ECL) sensors, colourimetric sensors and surface-enhanced Raman scattering (SERS) sensors. The ease of tunability, chemical stability, active sites including metals and functional groups, and porosity make MOFs an excellent precursor or composite material to yield excellent optical and electrical properties needed for practical sensing applications.

Microextraction of organophosphorus pesticides from food samples was achieved using a nitrogen-doped porous carbon derived from templated MOF [71]. The MOF in question is MIL-125, in which g-C_3N_4 was introduced as a template during the solvothermal synthesis process, creating g-C_3N_4@MOF composite. Furthermore, a porous carbon architecture was achieved by carbonization of the templated MOF, which in turn exhibited a large surface area and adsorption capacity, along with high stability. The sensor thus created was found to be successful in detecting organophosphorus pesticides (OPPs) from a variety of fruits and vegetable samples. A composite sensor derived from Ce-MOF as a sacrificial precursor is reported to be applied as an electrochemical sensor for detecting rutin [72]. Rutin is a common chemical with applications as an antioxidant, nutrition fortifier, and therapeutic agent against novel coronaviruses. Ce-MOF is prepared into rod-like powders using conventional methods. HCl and EDOT (3,4-ethylenedioxythiophene) are made into a solution and subsequent chemical treatment into Ce-MOF, synthesizing Ce-doped PEDOT (poly-EDOT). EDOT in itself is a conductive polymer used in various sensing applications. Overall, the Ce-doped PEDOT obtained using MOF as the sacrificial template performed as an excellent electrochemical sensor for detecting rutin from food samples. Yuwen Xu et al. synthesize a Ni-C compound from pyrolysis of hydrothermal Ni-MOF [73]. The resultant Ni-C composite was found to be an excellent electrochemical sensor for detecting histamine among food samples. Wei et al. prepared HNT/MIL-88 A and HNT/ZIF-67 composite (where HNT = halloysite nanotubes, MIL =Mat´eriaux de Institut Lavoisier, and ZIF =Zeolitic imidazolate frameworks) [74]. This composite has an active surface in which HNTs are grown by electrostatic adsorption, giving them appropriate properties for sensing applications. Fe_3O_4, when fabricated using MOF-808 (Fe_3O_4@MOF-808), provided an improved ability to capture nitrogen, oxyfluorfen, and bifenox from rice samples [75]. AuNPs (gold nanoparticles), when fabricated using UiO-66, were employed to detect amine and Sudan Red contaminants from chilli products [76]. Co/MnO@HC derived from MnCo-MOF-74 is reported to exhibit good electrochemical detection potential for detecting glucose in food samples [77]. In another example, Ag nanoparticles derived from MOFs were used as a superoxide anion radical sensor to detect antioxidants in food samples [78]. At times, the sensors in need of huge cavities are achieved by using bimetallic MOF precursors. Niu et al. designed a $NiCo_2O_4$@C/GCE compound

TABLE 8.2
MOF-derived materials for food pollution

Sensing material	Parent MOF	Contaminant	Food samples	Ref.
NiCo-LDH	NiCo-MOF	Hydrazine	Prawns, kelp, beer, soda	[80]
Fe-N-C nanozymes	Fe-ZIF-8/ Fe-ZIF-67	Organophosphorus pesticides (OPs)	Vegetables	[81]
CuO@C	Cu-BTB	Diflubenzuron	Tomato, cucumber	[82]
Carbon nanocomposites	M-MOF-NH2	Methyl parathion	Biosamples	[83]
Carbon nanocomposites	MOF 5	Herbicides	Drinking water	[84]
CuO@C$_3$N$_4$	Cu-MOF	H$_2$O$_2$	Fruits and Fish	[85]
Oxide nanoparticles	Cu-MOF	BPA	Biosamples and mineral water bottles	[86]
In$_2$O$_3$	MIL-68	Tetracycline	Animal products	[87]
Carbon nanocomposite	Cu-BTC MOF	Aflatoxin B1	Vegetable oil	[88]
AgNps/Co$_3$O$_4$@C	ZIF-9	Superoxide anion	Fetal calf serum	[89]

derived from Co/Ni-MOF to detect FZD (furazolidone) and CAP (chloramphenicol) in food samples such as milk and honey [79] (Table 8.2).

8.5 NOISE AND ELECTROMAGNETIC POLLUTION

Noise pollution accounts for the third significant environmental pollution. Some sources of noise pollution include aeroplanes, traffic, amplified music, mass transit, construction and industrial activity, emergency vehicles, garden or household types of equipment, public gatherings and technological devices. Sound-absorbing and attenuating materials are of significant interest when it comes to indoor and outdoor noise pollution. MOF-derived materials, albeit being less explored, are reported in several such noise remediation applications.

An ultra-thin MOF-based composite is reported to exhibit higher sound absorption compared to its constituent materials [90]. A Zr-based MOF (UiO-66) was integrated with a PVA solution and into a melamine foam, which is a commonly employed material. The resulting MOF is reported to have about 2.5 times higher attenuation properties than the original foam. The embedded Zr-MOF is said to act as a microscale Helmholtz resonator, improving sound attenuation. Another notable research includes the elastic aerogel from nanofibrous MOF, which is reported to have multifunctional abilities of low-frequency sound absorption and thermal insulation [91]. The synthesis route involves an electrospinning method to load MOF (ZIF-8) onto a polymer matrix (PAN) to create a PAN@ZIF-8 polymer composite. Furthermore, the PAN@ ZIF-8 electro-spun fibre membrane was cut into pieces and then dispersed into water/ tert-butanol by a high-speed homogenizer. Followed by a series of cross-linking to strengthen, the resultant PAN@ZIF-8 Kevlar was achieved. The PAN@ZIF-8 Kevlar is reported to have a sound absorption coefficient of 0.99 at 500 Hz and thermal conductivity close to air. ZIF-67, when grown in situ over wood (ZIF-67/wood), is used as a precursor to generate porous carbon magnetic composites, which possess good sound-insulating properties [92].

Electromagnetic pollution weighs in as the fifth major contributor to environmental pollution. Electromagnetic pollution can result from one or more electronic, electrical, industrial, or technological appliances. Some significant examples include WiFi routers, cell phones, network antennas, tracking devices, microwave ovens and other commonly used machines. Electromagnetic pollution is considered harmful to humans, birds, plants and other animals. It could lead to neurological diseases, cancer, infertility, and congenital defects, among other conditions. Electromagnetic radiation, in general, can be ionizing or non-ionizing radiation falling into a broad spectrum of frequencies (shortwave, radio, microwave, millimetre wave, infrared, visible light, and ultraviolet frequency). The field incorporating MOF-derived materials for suppressing electromagnetic pollution is vast and expanding. There have been numerous accounts of MOF-derived materials being explored for microwave suppression and other radiations. Some examples of MOF-derived materials used in the remediation of electromagnetic pollution are MOF-derived carbon composites [93–95], Ni/C composites [96], corn like MNO@C [97], and many others. Among these, doped carbons derived from MOFs in the forms of nanorods, flower-shaped, corn-shaped, yolk-shell arrangements, nanocages, nanocubes, etc., are being primarily explored.

8.6 GREENER MOF SYNTHESIS

While MOF-derived materials are being increasingly explored for environmental remediation, it becomes essential to make sure the MOF used in itself has less ecological footprint.

1. Choice of metals

   The central metal plays an integral role and will act as a significant component when the MOF decomposes. Hence, the choice of the central metal atom should be taken carefully to avoid possible environmental pollution. Alkaline earth metals (such as calcium and magnesium) are safer than transition metals (such as titanium, iron and manganese).

2. Choice of linkers

   Linkers tend to be the dominant reactive sites and thus need to be carefully chosen to avoid the formation of toxic byproducts that pollute the environment. Linkers such as carboxylic acids and imidazoles, when used in their protonated forms, form toxic byproducts such as HCl, H_2SO_4 and HNO_3. The use of biocompatible linkers might be preferable as they leave a less environmental footprint. For example, researchers have successfully used PET (Polyethylene terephthalate) plastic from waste as a green ligand for MOFs [98].

3. Choice of precursors

   The choice of precursors plays a vital role in deciding the cost of the end MOF product. However, the environmental footprint must be kept in mind during the selection of MOF precursors. For example, metal chlorides are attributed to causing severe corrosion due to acidic byproducts. Nitrates, chlorates, sulphates and acetates are hazardous compared to oxides and hydroxides. In general, precursors which produce less corrosive, oxidizing or toxic byproducts should be preferred.

4. Choice of solvents

   Hazardous solvents should be replaced with greener solvents in order to reduce their environmental impact. The most used yet toxic solvents include dimethylformamide (DMF), dimethylacetamide (DMAc), *N*-methylpyrrolidone (NMP), etc. Some comparatively green solvents include toluene, dimethylsulfoxide, xylene, *n*-butyl pyrrolidinone (substitute to NMP) and Dichloromethane. On the other hand, there would be much less environmental damage if solvents such as water, ionic liquids, acetic acid, acetone, methanol, and ethanol were employed. For example, a cellulose-derived solvent (cyrene) has been reported to be successfully used for several MOF membranes [99].

8.7 CONCLUSION

MOF-derived materials are being well-explored in various environmental applications. In the case of water pollution remediation, MOF-derived carbonaceous materials and MOF-derived magnetic materials are increasingly developed for the removal of challenging pollutant categories. When it comes to the removal of air pollutants, almost most of the MOF-derived materials employed are carbonaceous, such as activated carbon, activated carbon fibres, and carbon aerogel. Porous carbons have always been the widely employed material for air pollution control. MOFs, with their flexibility, robust surface, and architecture, give MOF-derived porous carbon materials an edge when battling air pollution. MOF-derived materials are seen as efficient in the microextraction of various harmful food pollutants. One of the underestimated pollution is from noise and electromagnetic radiation. Various MOF-derived materials and MOF-based composites are being used in this regard. It is essential to make sure the parent MOF synthesis procedures become greener, starting from the choice of metal, linkers, and precursors to solvents.

REFERENCES

1. Krupa, S. V.; Kickert, R. N. The Greenhouse Effect: Impacts of Ultraviolet-B (UV-B) Radiation, Carbon Dioxide (CO2), and Ozone (O3) on Vegetation. *Environ Pollut*, 1989, *61* (4). https://doi.org/10.1016/0269-7491(89)90166-8.
2. Peirce, J. J.; Weiner, R. F.; Vesilind, P. A. Nonpoint Source Water Pollution. In *Environmental Pollution and Control*; Butterworth-Heinemann, 1998, p 1. https://doi.org/10.1016/b978-075069899-3/50011-0.
3. Bhadra, B. N.; Yoo, D. K.; Jhung, S. H. Carbon-Derived from Metal-Organic Framework MOF-74: A Remarkable Adsorbent to Remove a Wide Range of Contaminants of Emerging Concern from Water. *Appl Surf Sci*, 2020, *504*. https://doi.org/10.1016/j.apsusc.2019.144348.
4. Zhang, Y.; Kang, X.; Guo, P.; Tan, H.; Zhang, S. H. Studies on the Removal of Phosphate in Water through Adsorption Using a Novel Zn-MOF and Its Derived Materials. *Arabian J Chem*, 2022, *15* (8). https://doi.org/10.1016/j.arabjc.2022.103955.

5. Song, X.; Chen, X.; Chen, W.; Ao, T. MOFs-Derived Fe, N-Co Doped Porous Carbon Anchored on Activated Carbon for Enhanced Phosphate Removal by Capacitive Deionization. *Sep Purif Technol*, 2023, *307*. https://doi.org/10.1016/j.seppur.2022.122694.

6. Ahmed, I.; Panja, T.; Khan, N. A.; Sarker, M.; Yu, J. S.; Jhung, S. H. Nitrogen-Doped Porous Carbons from Ionic Liquids@MOF: Remarkable Adsorbents for Both Aqueous and Nonaqueous Media. *ACS Appl Mater Interfaces*, 2017, *9* (11). https://doi.org/10.1021/acsami.7b00859.

7. del Rio, M.; Grimalt Escarabajal, J. C.; Turnes Palomino, G.; Palomino Cabello, C. Zinc/Iron Mixed-Metal MOF-74 Derived Magnetic Carbon Nanorods for the Enhanced Removal of Organic Pollutants from Water. *Chem Eng J*, 2022, *428*. https://doi.org/10.1016/j.cej.2021.131147.

8. Del Rio, M.; Turnes Palomino, G.; Palomino Cabello, C. Metal-Organic Framework@ Carbon Hybrid Magnetic Material as an Efficient Adsorbent for Pollutant Extraction. *ACS Appl Mater Interfaces*, 2020, *12* (5). https://doi.org/10.1021/acsami.9b19722.

9. Andrew Lin, K. Y.; Chang, H. A.; Chen, B. J. Multi-Functional MOF-Derived Magnetic Carbon Sponge. *J Mater Chem A Mater*, 2016, *4* (35). https://doi.org/10.1039/c6ta04619c.

10. Torad, N. L.; Hu, M.; Ishihara, S.; Sukegawa, H.; Belik, A. A.; Imura, M.; Ariga, K.; Sakka, Y.; Yamauchi, Y. Direct Synthesis of MOF-Derived Nanoporous Carbon with Magnetic Co Nanoparticles toward Efficient Water Treatment. *Small*, 2014, *10* (10). https://doi.org/10.1002/smll.201302910.

11. Jafari, Z.; Avargani, V. M.; Rahimi, M. R.; Mosleh, S. Magnetic Nanoparticles-Embedded Nitrogen-Doped Carbon Nanotube/Porous Carbon Hybrid Derived from a Metal-Organic Framework as a Highly Efficient Adsorbent for Selective Removal of Pb(II) Ions from Aqueous Solution. *J Mol Liq*, 2020, *318*. https://doi.org/10.1016/j.molliq.2020.113987.

12. Chen, D.; Chen, C.; Shen, W.; Quan, H.; Chen, S.; Xie, S.; Luo, X.; Guo, L. MOF-Derived Magnetic Porous Carbon-Based Sorbent: Synthesis, Characterization, and Adsorption Behavior of Organic Micropollutants. *Adv Powder Technol*, 2017, *28* (7). https://doi.org/10.1016/j.apt.2017.04.018.

13. Pan, Y.; Ding, Q.; Li, B.; Wang, X.; Liu, Y.; Chen, J.; Ke, F.; Liu, J. Self-Adjusted Bimetallic Zeolitic-Imidazolate Framework-Derived Hierarchical Magnetic Carbon Composites as Efficient Adsorbent for Optimizing Drug Contaminant Removal. *Chemosphere*, 2021, *263*. https://doi.org/10.1016/j.chemosphere.2020.128101.

14. Hasan, Z.; Cho, D. W.; Chon, C. M.; Yoon, K.; Song, H. Reduction of P-Nitrophenol by Magnetic Co-Carbon Composites Derived from Metal Organic Frameworks. *Chem Eng J*, 2016, *298*. https://doi.org/10.1016/j.cej.2016.04.029.

15. Wang, B.; Wang, C.; Yao, S.; Peng, Y.; Xu, Y. Plasma-Catalytic Degradation of Tetracycline Hydrochloride over Mn/γ-Al2O3 Catalysts in a Dielectric Barrier Discharge Reactor. *Plasma Sci Technol*, 2019, *21* (6). https://doi.org/10.1088/2058-6272/ab079c.

16. Liu, W. J.; Park, Y. K.; Bui, H. M.; Huy, N. N.; Lin, C. H.; Ghotekar, S.; Wi-Afedzi, T.; Lin, K. Y. A. Hofmann-MOF-Derived CoFeNi Nanoalloy@CNT as a Magnetic Activator for Peroxymonosulfate to Degrade Benzophenone-1 in Water. *J Alloys Compd*, 2023, *937*. https://doi.org/10.1016/j.jallcom.2022.165189.

17. Zhu, J.; Zhan, H.; Miao, X.; Zhao, K.; Zhou, Q. Terahertz Double-Exponential Model for Adsorption of Volatile Organic Compounds in Active Carbon. *J Phys D Appl Phys*, 2017, *50* (23). https://doi.org/10.1088/1361-6463/aa6e2f.

18. Gholidoust, A.; Atkinson, J. D.; Hashisho, Z. Enhancing CO2 Adsorption via Amine-Impregnated Activated Carbon from Oil Sands Coke. *Energy Fuels*, 2017, *31* (2). https://doi.org/10.1021/acs.energyfuels.6b02800.

19. Sjostrom, S.; Durham, M.; Bustard, C. J.; Martin, C. Activated Carbon Injection for Mercury Control: Overview. *Fuel*, 2010, *89* (6). https://doi.org/10.1016/j.fuel.2009.11.016.

20. Palliyarayil, A.; Saini, H.; Vinayakumar, K.; Selvarajan, P.; Vinu, A.; Kumar, N. S.; Sil, S. Advances in Porous Material Research towards the Management of Air Pollution. *Emergent Mater*, 2021. https://doi.org/10.1007/s42247-020-00151-9.

21. Mahmoodi, N. M.; Taghizadeh, M.; Taghizadeh, A. Activated Carbon/Metal-Organic Framework Composite as a Bio-Based Novel Green Adsorbent: Preparation and Mathematical Pollutant Removal Modeling. *J Mol Liq*, 2019, *277*. https://doi.org/10.1016/j.molliq.2018.12.050.

22. Askari, H.; Ghaedi, M.; Dashtian, K.; Azghandi, M. H. A. Rapid and High-Capacity Ultrasonic Assisted Adsorption of Ternary Toxic Anionic Dyes onto MOF-5-Activated Carbon: Artificial Neural Networks, Partial Least Squares, Desirability Function and Isotherm and Kinetic Study. *Ultrason Sonochem*, 2017, *37*. https://doi.org/10.1016/j.ultsonch.2016.10.029.

23. Kayal, S.; Chakraborty, A. Activated Carbon (Type Maxsorb-III) and MIL-101(Cr) Metal Organic Framework Based Composite Adsorbent for Higher CH4 Storage and CO2 Capture. *Chem Eng J*, 2018, *334*. https://doi.org/10.1016/j.cej.2017.10.080.

24. Shin, C.; Kim, K.; Choi, B. Deodorization Technology at Industrial Facilities Using Impregnated Activated Carbon Fiber. *J Chem Eng Jpn*, 2001, *34* (3). https://doi.org/10.1252/jcej.34.401.

25. Son, H. K.; Sivakumar, S.; Rood, M. J.; Kim, B. J. Electrothermal Adsorption and Desorption of Volatile Organic Compounds on Activated Carbon Fiber Cloth. *J Hazard Mater*, 2016, *301*. https://doi.org/10.1016/j.jhazmat.2015.08.040.

26. Ryu, D. Y.; Shimohara, T.; Nakabayashi, K.; Miyawaki, J.; Park, J. Il; Yoon, S. H. Urea/Nitric Acid Co-Impregnated Pitch-Based Activated Carbon Fiber for the Effective Removal of Formaldehyde. *J Ind Eng Chem*, 2019, *80*. https://doi.org/10.1016/j.jiec.2019.07.036.

27. Yang, S.; Zhu, Z.; Wei, F.; Yang, X. Enhancement of Formaldehyde Removal by Activated Carbon Fiber via in Situ Growth of Carbon Nanotubes. *Build Environ*, 2017, *126*. https://doi.org/10.1016/j.buildenv.2017.09.025.

28. Yue, Z.; Vakili, A. Activated Carbon–Carbon Composites Made of Pitch-Based Carbon Fibers and Phenolic Resin for Use of Adsorbents. *J Mater Sci*, 2017, *52* (21). https://doi.org/10.1007/s10853-017-1389-7.

29. Wang, C.; Yang, F.; Yang, W.; Ren, L.; Zhang, Y.; Jia, X.; Zhang, L.; Li, Y. PdO Nanoparticles Enhancing the Catalytic Activity of Pd/Carbon Nanotubes for 4-Nitrophenol Reduction. *RSC Adv*, 2015, *5* (35). https://doi.org/10.1039/c4ra16792a.

30. Baur, G. B.; Beswick, O.; Spring, J.; Yuranov, I.; Kiwi-Minsker, L. Activated Carbon Fibers for Efficient VOC Removal from Diluted Streams: The Role of Surface Functionalities. *Adsorption*, 2015, *21* (4). https://doi.org/10.1007/s10450-015-9667-7.

31. Attia, N. F.; Jung, M.; Park, J.; Jang, H.; Lee, K.; Oh, H. Flexible Nanoporous Activated Carbon Cloth for Achieving High H2, CH4, and CO2 Storage Capacities and Selective CO2/CH4 Separation. *Chem Eng J*, 2020, *379*. https://doi.org/10.1016/j.cej.2019.122367.

32. Li, M.; Lu, B.; Ke, Q. F.; Guo, Y. J.; Guo, Y. P. Synergetic Effect between Adsorption and Photodegradation on Nanostructured TiO2/Activated Carbon Fiber Felt Porous

Composites for Toluene Removal. *J Hazard Mater*, 2017, *333*. https://doi.org/10.1016/j.jhazmat.2017.03.019.

33. Palliyarayil, A.; Prakash, P. S.; Nandakumar, S.; Kumar, N. S.; Sil, S. Palladium Nanoparticles Impregnated Activated Carbon Material for Catalytic Oxidation of Carbon Monoxide. *Diam Relat Mater*, 2020, *107*. https://doi.org/10.1016/j.diamond.2020.107884.

34. Kang, S.; Hwang, J. Fabrication of Hollow Activated Carbon Nanofibers (HACNFs) Containing Manganese Oxide Catalyst for Toluene Removal via Two-Step Process of Electrospinning and Thermal Treatment. *Chem Eng J*, 2020, *379*. https://doi.org/10.1016/j.cej.2019.122315.

35. Ma, X.; Li, L.; Chen, R.; Wang, C.; Zhou, K.; Li, H. Doping of Alkali Metals in Carbon Frameworks for Enhancing CO2 Capture: A Theoretical Study. *Fuel*, 2019, *236*. https://doi.org/10.1016/j.fuel.2018.08.166.

36. Hu, J.; Liu, Y.; Liu, J.; Gu, C. Computational Screening of Alkali, Alkaline Earth, and Transition Metals Alkoxide-Functionalized Metal-Organic Frameworks for CO2 Capture. *J Phys Chem C*, 2018, *122* (33). https://doi.org/10.1021/acs.jpcc.8b05334.

37. Gan, G.; Li, X.; Fan, S.; Wang, L.; Qin, M.; Yin, Z.; Chen, G. Carbon Aerogels for Environmental Clean-Up. *Eur J Inorg Chem*, 2019. https://doi.org/10.1002/ejic.201801512.

38. Thubsuang, U.; Sukanan, D.; Sahasithiwat, S.; Wongkasemjit, S.; Chaisuwan, T. Highly Sensitive Room Temperature Organic Vapor Sensor Based on Polybenzoxazine-Derived Carbon Aerogel Thin Film Composite. *Mater Sci Eng: B*, 2015, *200*. https://doi.org/10.1016/j.mseb.2015.06.010.

39. Mohamad Nor, N.; Lau, L. C.; Lee, K. T.; Mohamed, A. R. Synthesis of Activated Carbon from Lignocellulosic Biomass and Its Applications in Air Pollution Control – A Review. *J Environ Chem Eng*, 2013. https://doi.org/10.1016/j.jece.2013.09.017.

40. Lee, Y. W.; Kim, H. J.; Park, J. W.; Choi, B. U.; Choi, D. K.; Park, J. W. Adsorption and Reaction Behavior for the Simultaneous Adsorption of NO-NO2 and SO2 on Activated Carbon Impregnated with KOH. *Carbon N Y*, 2003, *41* (10). https://doi.org/10.1016/S0008-6223(03)00105-2.

41. Brady, T. A.; Rostam-Abadi, M.; Rood, M. J. Applications for Activated Carbons from Waste Tires: Natural Gas Storage and Air Pollution Control. *Gas Separation and Purification*, 1996, *10* (2). https://doi.org/10.1016/0950-4214(96)00007-2.

42. Mui, E. L. K.; Ko, D. C. K.; McKay, G. Production of Active Carbons from Waste Tyres – A Review. *Carbon*, 2004. https://doi.org/10.1016/j.carbon.2004.06.023.

43. Khajavi, R.; Bahadoran, M. M. S.; Bahador, A.; Khosravi, A. Removal of Microbes and Air Pollutants Passing through Nonwoven Polypropylene Filters by Activated Carbon and Nanosilver Colloidal Layers. *J Ind Text*, 2013, *42* (3). https://doi.org/10.1177/1528083711434653.

44. Abumaizar, R. J.; Kocher, W.; Smith, E. H. Biofiltration of BTEX Contaminated Air Streams Using Compost-Activated Carbon Filter Media. *J Hazard Mater*, 1998, *60* (2). https://doi.org/10.1016/S0304-3894(97)00046-0.

45. Ao, C. H.; Lee, S. C. Indoor Air Purification by Photocatalyst TiO2 Immobilized on an Activated Carbon Filter Installed in an Air Cleaner. *Chem Eng Sci*, 2005, *60* (1). https://doi.org/10.1016/j.ces.2004.01.073.

46. Khayan, K.; Anwar, T.; Wardoyo, S.; Puspita, W. L. Respiratory Mask Using a Combination of Spunbond, Meltblown, and Activated Carbon Materials for Reducing Exposure to CO: An In Vivo Study. *Environ Sci Pollut Res*, 2021, *28* (15). https://doi.org/10.1007/s11356-020-09476-8.

47. Gallego, E.; Roca, F. J.; Perales, J. F.; Guardino, X. Experimental Evaluation of VOC Removal Efficiency of a Coconut Shell Activated Carbon Filter for Indoor Air Quality Enhancement. *Build Environ*, 2013, *67*. https://doi.org/10.1016/j.buildenv.2013.05.003.

48. An, Y.; Fu, Q.; Zhang, D.; Wang, Y.; Tang, Z. Performance Evaluation of Activated Carbon with Different Pore Sizes and Functional Groups for VOC Adsorption by Molecular Simulation. *Chemosphere*, 2019, *227*. https://doi.org/10.1016/j.chemosphere.2019.04.011.

49. Metts, T. A.; Batterman, S. A. Effect of VOC Loading on the Ozone Removal Efficiency of Activated Carbon Filters. *Chemosphere*, 2006, *62* (1). https://doi.org/10.1016/j.chemosphere.2005.04.049.

50. Leson, G.; Winer, A. M. Biofiltration: An Innovative Air Pollution Control Technology for Voc Emissions. *J Air Waste Manage Assoc*, 1991, *41* (8). https://doi.org/10.1080/10473289.1991.10466898.

51. Shafeeyan, M. S.; Daud, W. M. A. W.; Houshmand, A.; Shamiri, A. A Review on Surface Modification of Activated Carbon for Carbon Dioxide Adsorption. *J Anal Appl Pyrolysis*, 2010. https://doi.org/10.1016/j.jaap.2010.07.006.

52. Rashidi, N. A.; Yusup, S.; Loong, L. H. Kinetic Studies on Carbon Dioxide Capture Using Activated Carbon. *Chemical Engineering Transactions*, 2013; *35*, 361–366. https://doi.org/10.3303/CET1335060.

53. Mohd Azmi, N. Z.; Buthiyappan, A.; Abdul Raman, A. A.; Abdul Patah, M. F.; Sufian, S. Recent Advances in Biomass Based Activated Carbon for Carbon Dioxide Capture – A Review. *J Ind Eng Chem*, 2022. https://doi.org/10.1016/j.jiec.2022.08.021.

54. Ahmed, M. B.; Hasan Johir, M. A.; Zhou, J. L.; Ngo, H. H.; Nghiem, L. D.; Richardson, C.; Moni, M. A.; Bryant, M. R. Activated Carbon Preparation from Biomass Feedstock: Clean Production and Carbon Dioxide Adsorption. *J Clean Prod*, 2019, *225*. https://doi.org/10.1016/j.jclepro.2019.03.342.

55. Sidheswaran, M. A.; Destaillats, H.; Sullivan, D. P.; Cohn, S.; Fisk, W. J. Energy Efficient Indoor VOC Air Cleaning with Activated Carbon Fiber (ACF) Filters. *Build Environ*, 2012, *47* (1). https://doi.org/10.1016/j.buildenv.2011.07.002.

56. Yang, S.; Zhu, Z.; Wei, F.; Yang, X. Carbon Nanotubes/Activated Carbon Fiber Based Air Filter Media for Simultaneous Removal of Particulate Matter and Ozone. *Build Environ*, 2017, *125*. https://doi.org/10.1016/j.buildenv.2017.08.040.

57. Yao, M.; Zhang, Q.; Hand, D. W.; Perram, D.; Taylor, R. Adsorption and Regeneration on Activated Carbon Fiber Cloth for Volatile Organic Compounds at Indoor Concentration Levels. *J Air Waste Manage Assoc*, 2009, *59* (1). https://doi.org/10.3155/1047-3289.59.1.31.

58. Hashisho, Z.; Rood, M.; Botich, L. Microwave-Swing Adsorption to Capture and Recover Vapors from Air Streams with Activated Carbon Fiber Cloth. *Environ Sci Technol*, 2005, *39* (17). https://doi.org/10.1021/es050338z.

59. Le Cloirec, P. Adsorption onto Activated Carbon Fiber Cloth and Electrothermal Desorption of Volatile Organic Compound (VOCs): A Specific Review. *Chin J Chem Eng*, 2012, *20* (3). https://doi.org/10.1016/S1004-9541(11)60207-3.

60. Rathore, R. S.; Srivastava, D. K.; Agarwal, A. K.; Verma, N. Development of Surface Functionalized Activated Carbon Fiber for Control of NO and Particulate Matter. *J Hazard Mater*, 2010, *173* (1–3). https://doi.org/10.1016/j.jhazmat.2009.08.071.

61. Park, S.; Yaqub, M.; Lee, S.; Lee, W. Adsorption of Acetaldehyde from Air by Activated Carbon and Carbon Fibers. *Environ Eng Res*, 2022, *27* (2). https://doi.org/10.4491/eer.2020.549.

62. Rehman, A. ur; Baek, J. W.; Rene, E. R.; Sergienko, N.; Behera, S. K.; Park, H. S. Effect of Process Parameters Influencing the Chemical Modification of Activated Carbon Fiber for Carbon Dioxide Removal. *Process Saf Environ Prot*, 2018, *118*. https://doi.org/10.1016/j.psep.2018.07.004.

63. Lee, J. H.; Lee, S. H.; Suh, D. H. Utilization of Carbon Dioxide onto Activated Carbon Fibers for Surface Modification. *Carbon Lett*, 2020, *30* (1). https://doi.org/10.1007/s42823-019-00076-2.

64. Zhang, X.; He, W.; Zhang, R.; Wang, Q.; Liang, P.; Huang, X.; Logan, B. E.; Fellinger, T. P. High-Performance Carbon Aerogel Air Cathodes for Microbial Fuel Cells. *ChemSusChem*, 2016, *9* (19). https://doi.org/10.1002/cssc.201600590.

65. Jatoi, A. S.; Hashmi, Z.; Mazari, S. A.; Abro, R.; Sabzoi, N. Recent Developments and Progress of Aerogel Assisted Environmental Remediation: A Review. *J Porous Mater*, 2021, *28* (6). https://doi.org/10.1007/s10934-021-01136-7.

66. Han, S.; Sun, Q.; Zheng, H.; Li, J.; Jin, C. Green and Facile Fabrication of Carbon Aerogels from Cellulose-Based Waste Newspaper for Solving Organic Pollution. *Carbohydr Polym*, 2016, *136*. https://doi.org/10.1016/j.carbpol.2015.09.024.

67. Ding, C.; Liu, Y.; Xie, P.; Lan, J.; Yu, Y.; Fu, X.; Yang, X.; Zhong, W. H. A Novel Carbon Aerogel Enabling Respiratory Monitoring for Bio-Facial Masks. *J Mater Chem A Mater*, 2021, *9* (22). https://doi.org/10.1039/d1ta00794g.

68. Hsan, N.; Dutta, P. K.; Kumar, S.; Bera, R.; Das, N. Chitosan Grafted Graphene Oxide Aerogel: Synthesis, Characterization and Carbon Dioxide Capture Study. *Int J Biol Macromol*, 2019, *125*. https://doi.org/10.1016/j.ijbiomac.2018.12.071.

69. Ello, A. S.; Yapo, J. A.; Trokourey, A. N-Doped Carbon Aerogels for Carbon Dioxide (CO_2) Capture. *Afr J Pure Appl Chem*, 2013, *7* (2), 61–66.

70. Saeed, A. M.; Rewatkar, P. M.; Majedi Far, H.; Taghvaee, T.; Donthula, S.; Mandal, C.; Sotiriou-Leventis, C.; Leventis, N. Selective CO2 Sequestration with Monolithic Bimodal Micro/Macroporous Carbon Aerogels Derived from Stepwise Pyrolytic Decomposition of Polyamide-Polyimide-Polyurea Random Copolymers. *ACS Appl Mater Interfaces*, 2017, *9* (15). https://doi.org/10.1021/acsami.7b01910.

71. Pang, Y.; Zang, X.; Li, H.; Liu, J.; Chang, Q.; Zhang, S.; Wang, C.; Wang, Z. Solid-Phase Microextraction of Organophosphorous Pesticides from Food Samples with a Nitrogen-Doped Porous Carbon Derived from g-C3N4 Templated MOF as the Fiber Coating. *J Hazard Mater*, 2020, *384*. https://doi.org/10.1016/j.jhazmat.2019.121430.

72. Wang, Y.; Chen, J.; Wang, C.; Zhang, L.; Yang, Y.; Chen, C.; Xie, Y.; Zhao, P.; Fei, J. An Electrochemical Sensor Based on Ce-MOF-Derived Ce-Doped Poly(3,4-Ethylenedioxythiophene) Composite for Efficient Determination of Rutin in Food. *Talanta*, 2023, *263*. https://doi.org/10.1016/j.talanta.2023.124678.

73. Xu, Y.; Cheng, Y.; Jia, Y.; Ye, B. C. Synthesis of MOF-Derived Ni@C Materials for the Electrochemical Detection of Histamine. *Talanta*, 2020, *219*. https://doi.org/10.1016/j.talanta.2020.121360.

74. Wei, Y.; Liu, R.; Liu, H.; Sun, Y.; Liu, G.; Wang, Z.; Luo, Y.; Tang, X.; Zhang, X.; Hu, J. Controllable One-Dimensional Growth of Metal-Organic Frameworks Based on Uncarved Halloysite Nanotubes as High-Efficiency Solar-Fenton Catalysts. *J Phys Chem C*, 2021, *125* (46). https://doi.org/10.1021/acs.jpcc.1c08054.

75. Meng, Y.; Yeqing, J.; Peiru, Q.; Yahui, W.; Qianqian, J.; Manman, W.; Qian, W.; Yulan, H. Determination of Three Diphenyl Ether Herbicides in Rice by Magnetic Solid Phase Extraction Using Fe3O4 @MOF-808 Coupled with High Performance Liquid Chromatography. *Chin J Chromatogr (Se Pu)*, 2021, *39* (3). https://doi.org/10.3724/SP.J.1123.2020.06007.

76. Fu, J.; Lai, H.; Zhang, Z.; Li, G. UiO-66 Metal-Organic Frameworks/Gold Nanoparticles Based Substrates for SERS Analysis of Food Samples. *Anal Chim Acta*, 2021, *1161*. https://doi.org/10.1016/j.aca.2021.338464.

77. Zhang, Y.; Huang, Y.; Gao, P.; Yin, W.; Yin, M.; Pu, H.; Sun, Q.; Liang, X.; Fa, H. bao. Bimetal-Organic Frameworks MnCo-MOF-74 Derived Co/MnO@HC for the Construction of a Novel Enzyme-Free Glucose Sensor. *Microchem J*, 2022, *175*. https://doi.org/10.1016/j.microc.2021.107097.

78. Jiang, X.; Liu, X.; Wu, T.; Li, L.; Zhang, R.; Lu, X. Metal–Organic Framework Derived Carbon-Based Sensor for Monitoring of the Oxidative Stress of Living Cell and Assessment of Antioxidant Activity of Food Extracts. *Talanta*, 2019, *194*. https://doi.org/10.1016/j.talanta.2018.10.093.

79. Niu, X.; Bo, X.; Guo, L. MOF-Derived Hollow NiCo2O4/C Composite for Simultaneous Electrochemical Determination of Furazolidone and Chloramphenicol in Milk and Honey. *Food Chem*, 2021, *364*. https://doi.org/10.1016/j.foodchem.2021.130368.

80. Tang, X.; Zhao, S.; Wu, J.; He, Z.; Zhang, Y.; Huang, K.; Zou, Z.; Xiong, X. Construction of Rose Flower-Like NiCo-LDH Electrode Derived from Bimetallic MOF for Highly Sensitive Electrochemical Sensing of Hydrazine in Food Samples. *Food Chem*, 2023, *427*. https://doi.org/10.1016/j.foodchem.2023.136648.

81. Shen, Z.; Xu, D.; Wang, G.; Geng, L.; Xu, R.; Wang, G.; Guo, Y.; Sun, X. Novel Colorimetric Aptasensor Based on MOF-Derived Materials and Its Applications for Organophosphorus Pesticides Determination. *J Hazard Mater*, 2022, *440*. https://doi.org/10.1016/j.jhazmat.2022.129707.

82. Ji, X. X.; Liu, Y. L.; Chang, X. Y.; Li, R. L.; Ye, F.; Yang, L.; Fu, Y. An Electrochemical Sensor Derived from Cu-BTB MOF for the Efficient Detection of Diflubenzuron in Food and Environmental Samples. *Food Chem*, 2023, *428*. https://doi.org/10.1016/j.foodchem.2023.136802.

83. Dong, S.; Peng, L.; Wei, W.; Huang, T. Three MOF-Templated Carbon Nanocomposites for Potential Platforms of Enzyme Immobilization with Improved Electrochemical Performance. *ACS Appl Mater Interfaces*, 2018, *10* (17). https://doi.org/10.1021/acsami.8b00702.

84. Liu, C.; Wang, P.; Liu, X.; Yi, X.; Zhou, Z.; Liu, D. Multifunctional β-Cyclodextrin MOF-Derived Porous Carbon as Efficient Herbicides Adsorbent and Potassium Fertilizer. *ACS Sustain Chem Eng*, 2019, *7* (17). https://doi.org/10.1021/acssuschemeng.9b01911.

85. Lu, M.; Wang, Z.; Xie, W.; Zhang, Z.; Su, L.; Chen, Z.; Xiong, Y. Correction to: Cu-MOF Derived CuO@g-C3N4 Nanozyme for Cascade Catalytic Colorimetric Sensing. *Anal Bioanal Chem*, 2023, *415* (24). https://doi.org/10.1007/s00216-023-04892-4.

86. Saravanakumar, V.; Rajagopal, V.; Narayanan, K.; Nesakumar, N.; Kathiresan, M.; Suryanarayanan, V.; Anandan, S. Cu-MOF/Pd Derived Oxide Nanoparticles Based Carbon Composite – An Innovative Electrochemical Sensing Platform for Bisphenol A. *J Alloys Compd*, 2023, *963*. https://doi.org/10.1016/j.jallcom.2023.171216.

87. Feng, Y.; Yan, T.; Wu, T.; Zhang, N.; Yang, Q.; Sun, M.; Yan, L.; Du, B.; Wei, Q. A Label-Free Photoelectrochemical Aptasensing Platform Base on Plasmon Au Coupling with MOF-Derived In2O3@g-C3N4 Nanoarchitectures for Tetracycline Detection. *Sens Actuators B Chem*, 2019, *298*. https://doi.org/10.1016/j.snb.2019.126817.

88. Ma, F.; Cai, X.; Mao, J.; Yu, L.; Li, P. Adsorptive Removal of Aflatoxin B1 from Vegetable Oils via Novel Adsorbents Derived from a Metal-Organic Framework. *J Hazard Mater*, 2021, *412*. https://doi.org/10.1016/j.jhazmat.2021.125170.

89. Jiang, X.; Xiuhui, L.; Tiaodi, W.; Lin, L.; Rongjin, Z.; and Xiaoquan, L. Metal–Organic Framework Derived Carbon-Based Sensor for monitoring of the oxidative stress of living cell and assessment of antioxidant activity of food extracts. *Talanta*, 2019, *194*, 591–597.

90. Lan, Y. C.; Kamal, S.; Lin, C. C.; Liu, Y. H.; Lu, K. L. Ultra-Thin Zr-MOF/PVA/Melamine Composites with Remarkable Sound Attenuation Effects. *Microporous and Mesoporous Materials*, 2023, *360*. https://doi.org/10.1016/j.micromeso.2023.112668.

91. Fan, S. T.; Zhang, Y.; Tan, M.; Wang, J. X.; Huang, C. Y.; Li, B. J.; Zhang, S. Multifunctional Elastic Aerogels of Nanofibrous Metal–Organic Framework for Thermal Insulation and Broadband Low-Frequency Sound Absorption. *Compos Sci Technol*, 2023, *242*. https://doi.org/10.1016/j.compscitech.2023.110183.

92. Ma, X.; Guo, H.; Zhang, C.; Chen, D.; Tian, Z.; Wang, Y.; Chen, Y.; Wang, S.; Han, J.; Lou, Z.; et al. ZIF-67/Wood Derived Self-Supported Carbon Composites for Electromagnetic Interference Shielding and Sound and Heat Insulation. *Inorg Chem Front*, 2022, *14*. https://doi.org/10.1039/d2qi01943d.

93. Zhang, X.; Qiao, J.; Jiang, Y.; Wang, F.; Tian, X.; Wang, Z.; Wu, L.; Liu, W.; Liu, J. Carbon-Based MOF Derivatives: Emerging Efficient Electromagnetic Wave Absorption Agents. *Nano-Micro Letters*, 2021. https://doi.org/10.1007/s40820-021-00658-8.

94. Liu, W.; Liu, L.; Yang, Z.; Xu, J.; Hou, Y.; Ji, G. A Versatile Route toward the Electromagnetic Functionalization of Metal-Organic Framework-Derived Three-Dimensional Nanoporous Carbon Composites. *ACS Appl Mater Interfaces*, 2018, *10* (10). https://doi.org/10.1021/acsami.8b00320.

95. Ren, Y.; Wang, X.; Ma, J.; Zheng, Q.; Wang, L.; Jiang, W. Metal-Organic Framework-Derived Carbon-Based Composites for Electromagnetic Wave Absorption: Dimension Design and Morphology Regulation. *J Mater Sci Technol*, 2023, *132*. https://doi.org/10.1016/j.jmst.2022.06.013.

96. Wu, N.; Zhao, B.; Liu, J.; Li, Y.; Chen, Y.; Chen, L.; Wang, M.; Guo, Z. MOF-Derived Porous Hollow Ni/C Composites with Optimized Impedance Matching as Lightweight Microwave Absorption Materials. *Adv Compos Hybrid Mater*, 2021, *4* (3). https://doi.org/10.1007/s42114-021-00307-z.

97. Shen, Z.; Yang, H.; Liu, C.; Guo, E.; Huang, S.; Xiong, Z. Polymetallic MOF-Derived Corn-Like Composites for Magnetic-Dielectric Balance to Facilitate Broadband Electromagnetic Wave Absorption. *Carbon N Y*, 2021, *185*. https://doi.org/10.1016/j.carbon.2021.09.041.

98. Huang, Y. T.; Lai, Y. L.; Lin, C. H.; Wang, S. L. Direct Use of Waste PET as Unfailing Source of Organic Reagents in the Synthesis of Intrinsic White/Yellow Luminescent Nanoporous Zincophosphates. *Green Chem*, 2011, *13* (8). https://doi.org/10.1039/c1g c15427c.

99. Zhang, J.; White, G. B.; Ryan, M. D.; Hunt, A. J.; Katz, M. J. Dihydrolevoglucosenone (Cyrene) as a Green Alternative to N,N-Dimethylformamide (DMF) in MOF Synthesis. *ACS Sustain Chem Eng*, 2016, *4* (12). https://doi.org/10.1021/acssuschem eng.6b02115.

9 Emerging Applications

9.1 INTRODUCTION

Well-defined three-dimensional porous MOFs constructed from organic linkers and metal, or metal oxide nodes hold significant potential across a spectrum of applications, extending beyond sensing, energy, catalysis, gas adsorption, storage, and separation applications. The pore size of the MOF material allows storage of small molecules like H_2, CO_2, and CH_4. However, the pore size of the MOF materials used in biomedical applications should be large enough to allow larger biomolecules such as proteins. Furthermore, the loading of the voids of the MOF materials with different nanomaterials such as metal nanoclusters, metal nanoparticles, large molecules, and proteins, imparts the material the excellent properties required for applications including controlled drug delivery, plasmonic, and battery materials. Controlled loading of nanoparticles in the voids of MOF material is challenging. The liquid phase epitaxy (LPE) process facilitates the design of novel nanomaterials by enabling the controlled loading of nanoparticles in the framework and generating the freestanding heterocrystals on suitable substrates through hetero-epitaxial growth of MOF-on-MOF thin films [1]. However, these MOFs mounted on suitable substrates (SURMOF) have the major limitation of the presence of cytotoxic metal ions and poor water stability. Addressing these challenges has been the research focus of many researchers. In recent days, there has been significant attention on exploring the potential of MOF-derived designer materials in the realms of enzyme-catalytic reactions and various biomedical applications.

9.2 NANOZYMES

Nanozymes, the artificial enzymes, are emerging functional nanomaterials with catalytic behaviour similar to natural enzymes developed to overcome the limitations of natural enzymes such as their denaturation behaviour, the difficulty of recycling, and high cost. Other than these, nanozymes have various advantages over natural or conventional artificial enzymes such as ease of mass production, long-term storage, tunable activity, desirable properties through tuning of size, shape, and composition of nanozymes, multifunction, etc. MOFs can take various forms depending on the choice of metal precursor and organic linker, influencing their size, shape, structure,

DOI: 10.1201/9781003432357-11

and enzymatic catalytic behaviour. The different functional nanomaterials including carbon-based, metal-based, metal oxide-based, metal-organic framework-based, metal chalcogenide-based, and metal hydroxide-based nanomaterials, exhibit enzymatic catalytic behaviour. MOF can be used as a template or precursor to derive the different nanozyme materials. The different types of MOF-based nanozymes include pristine MOFs, modified MOFs, enzyme-encapsulated MOF composites, and MOF derivatives [2]. The MOF derivatives as nanozymes have the advantages of better stability, availability of abundant active sites, and tunable porous framework structure originated from the parent MOF that can be controlled through the choice of metal precursor and organic bridging molecule. The MOF derivatives including porous carbon-based and metal oxide-based nanozymes can be obtained by pyrolysis and etching of MOF materials.

Pyrolysis of the MOF material is the most common way of obtaining the MOF-derived nanozymes which converts the original structure of MOF with metal nodes bridged by organic ligands into the uniformly distributed metal oxide nanoparticles with porous carbonaceous structure, respectively. Metal oxide nanoparticles as nanozymes have the limitations of aggregation due to their high surface energy which can be prevented when it is derived from a metal-organic framework self-sacrificial precursor due to the presence of porous carbonaceous structure that helps in preventing the aggregation of metal oxide nanoparticles by keeping them separated. Cerium containing MOF was used as a sacrificed precursor to obtain the ultrasmall nanoparticles of CeO_2 (n-CeO_2 NSs and a-CeO_2 NSs) uniformly dispersed in a porous carbonaceous support structure on calcination at 400°C in nitrogen and air atmosphere, respectively. Oxidation of 3,3',5,5'-tetramethylbenzidine is reported to be the highest with n-CeO_2 compared to the other forms of CeO_2-based nanozymes due to the intrinsic porous structure which enhances the mass transport process during the catalytic oxidation process [3]. The therapeutic effect of n-CeO_2 NSs is evidenced by their ability to be used as the drug carrier for the anticancer drug DOX. Despite reduced active sites on loading the DOX drug on n-CeO_2 carriers, the unaffected Adenosine 5'-triphosphate disodium salt hydrate ATP deprivation evidences the recovered enzymatic activity of DOX@n-CeO_2 and the enhanced anticancer effect.

Li et al. derived homogeneous bimetallic oxide dispersed within the hollow nanocages (HNC) through low-temperature calcination of bimetallic MOF material which was obtained by ion assistant solvothermal process. Among the different secondary metals added to the CoM-HNC MOF structure (where M=Ni, Mn, Cu, and Zn), the Cu doping followed by subsequent low-temperature calcination demonstrated the highest oxidase-like activity as well as peroxymonosulfate (PMS) activator. The PMS-activating behaviour exhibited by C-CoCu-HNC makes them a suitable candidate for removing organic pollutants like rhodamine B, while the oxidase-like behaviour renders them a suitable application in medical diagnostics [4]. The direct pyrolysis of Mg-based MOF material resulted in a graphene-based Mg-centred cofactor catalyst. The design of the catalyst by mimicking the Mg cofactor configuration has practical difficulties due to their instability in the experimental environment. However, the coordination of magnesium with nitrogen-heterocyclic molecules allows for tuning the energy level of the highest occupied molecular orbital (HOMO) of Mg which provides active sites for the oxygen reduction reaction [5]. Wang et al coated ZIF-8 precursor with mesoporous silica. They subsequently subjected the mesoporous-coated ZIF-8 precursor to surface-protected pyrolysis followed by the removal of mesoporous silica by NaOH etching. This resulted in carbon nanospheres containing porphyrin-like zinc centres, referred to as PMCS. The mesoporous coating acts as a physical barrier to the aggregation of carbon particles during the pyrolysis process. The PMCS particles were employed in cancer therapy. PEGylated PMCS particles demonstrated excellent in-vitro and in-vivo biocompatibility as well as tumour cell necrosis/apoptosis upon laser irradiation [6]. Fe-N-C single atom nanozyme (Fe-SAzyme) was obtained by the pyrolysis of $Fe(acac)3@ZIF-8$ NPs at 800°C for 3 h with a heating rate of 3°C/min and the subsequent washing with 0.05M H_2SO_4 for 24 h to remove the Fe nanoparticles and oxides formed during the pyrolysis. On carbonization, the organic linker of ZIF-8 results in the formation of amorphous carbon, while the $Fe(acac)_3$ leads to the formation of atomic ions coordinated with the adjacent nitrogen atoms available with the organic linker of ZIF-8. Fe-SAzyme was further employed for its peroxidase-like catalytic activity in the oxidation of galactose which produces H_2O_2. The catalytic activity of Fe-SAzyme on H_2O_2 forms hydroxyl radicals which helps in the catalytic oxidization of 3,3',5,5'-tetramethylbenzidine (TMB). The TMB/Fe-SAzyme/galactose/galactose oxidase system quantitatively detected the galactose by the colourimetric method in the presence of galactose oxidase with a linear dynamic range of 50–500 μM with the corresponding limit of detection (LOD) of 10 μM [7].

MOF material can also be used as a support for the single-atom enzymes that are reported in targeted tumour cell treatment. The tumour cells are targeted by photodynamic therapy (PDT), one of the most efficient methods for targeting tumour cells with remarkably reduced side effects. PDT depends on the formation of molecular oxygen which produces highly reactive singlet oxygen that exhibits potent cytotoxicity against the tumour cells. The formation of molecular oxygen and the subsequent action of the singlet oxygen on tumour cells can be controlled precisely by localizing the photoirradiation through adept temporal and spatial control. However, the microvascular damage that PDT could cause can prevent the transport of molecular oxygen which will worsen the tumour hypoxia. A promising solution lies in the utilization of a single-atom catalyst anchored on a solid support to the generation of molecular oxygen directly at the tumour sites. Compared to conventional catalyst supports, MOF supports have the advantages of a well-defined coordination network and tunable pore size. Wang et al. synthesized a single-atom Ru catalyst by incorporating Ru atoms into the $Mn_3[Co(CN)_6]_2$ framework where Ru partially replaces Co and serves as the single-atom catalytic site for localized oxygen generation. The single-atom enzyme (OxgeMCC-r SAE) was obtained by encapsulating organic ligand, metal ions, and the photosensitizer (chlorin e6 (Ce6)) by polyvinylpyrrolidone. The high porosity of MOF support material allows substantial loading capacity of the photosensitizer, while the commendable catalytic ability and catalytic durability arise from six unsaturated coordination between Ru and carbon through the replacement of Co ions by catalytic Ru in the framework, which in the rapid onsite generation of oxygen without self-consumption and any external activation [8].

Not only the MOF derivatives obtained by the pyrolysis strategy but the MOF material as such due to their inherent Lewis acid and base sites available in their structure exhibits intrinsic enzyme-mimicking activities. Further post-synthetic modification helps in tailoring the catalytic properties of the MOF material. Amino functionalization of UiO-66 resulted in a 20-fold enhancement in phosphate-ester hydrolysis rate over the parent MOF material which can be attributed to the amino groups which act as a proton transfer agent. The amino groups will act as Bronsted base sites initially, and then transition into Bronsted acid sites during the catalytic cycle [9]. Similarly, ZIF-8 was modified to replace the parent organic linker, 2-methyl imidazolate, with unsubstituted imidazolate to obtain the MOF material with a large opening without obstruction of the substituents in the organic linker which helps in enhancing the surface area. Unlike the parent ZIF-8 of single aperture type, the linker-modified ZIF-8 exhibits apertures containing four-linker and six-linker rings. This was further employed in Bronsted base catalysis of n-butyllithium [10]. Post-synthetic modification of ZIF-67, reported by Sang et al., involves loading of 3-amino-1,2,4-triazole (3-AT) in ZIF-8 that forms an amorphous coating by the coordination of Co^{2+} ions and 3-AT resulting in reduced surface area from 1183.3 m^2/g for ZIF-67 nanoparticles to 172.4 m^2/g for ZIF-67-AT. Subsequently, it was functionalized with polyethylene glycol (PEG) to obtain the disruptor PZIF-67-AT rendering it biocompatible and stable under physiological conditions. The superoxide dismutase (SOD) like activity of the disruptor molecule, PZIF-67-AT, helped in the increased formation of H_2O_2 molecules and controls the scavenging of H_2O_2 by catalase and glutathione that causes accumulation of H_2O_2 in the cancer cells. The elevated H_2O_2 level in the cancer cells generates the ·OH that induces oxidative stress and strengthens the Fenton-reaction-based chemodynamic therapy [11].

9.3 MOF-POLYMER-BASED MATERIALS FOR BIOMEDICAL APPLICATIONS

MOFs, on the one hand, possess excellent porous architecture yet exhibit brittleness. Polymers, on another hand, possess flexibility and malleability. The mix of the advantages of these systems brings in flexible porous composites useful for many applications. There have been interesting reports, especially on biomedical applications including biosensors and drug delivery. One method to achieve such a system is to utilize MOFs as an initial template for polymer generation. Usually, polymer synthesis is performed using inorganic templates such as porous silica or zeolites. While the removal of such hard template requires special conditions, MOFs are known to be removed by mild conditions. This provides an easier way to synthesize useful polymer membranes without losing the superior porous architecture derived from the MOF precursors. In general, MOF-polmer composites can be achieved through various methods such as polymer-grating into MOF surface, polymerization of monomers within MOFs, polymer-templating MOFs, as well as MOFs templating polymers.

MOF-polymer composites are explored for several biosensing applications. Biosensors are analytical devices that utilize the bioreceptor that interacts with various analytes to produce measurable output. Biosensors can be of use in a variety of biomedical research including antibody interaction, protein docking studies, enzymatic interactions, nucleic acid interactions, and microbial detectors. For example, an MOF (NH_2-MIL-88B) integrated with a polymer (poly-styrene-acrylic acid) is reported to form a multilayered photonic crystal (PC) sensor [12]. The resulting stimuli response can be helpful in detecting various exposures such as those of benzene, toluene, ethylbenzene, and BTEX (xylene). A composite of Cu-MOF with a conducting polymer (polyaniline) is further modified to create an immunosensor which is found to be successful in detecting a crucial cardiac marker (troponin I/cTnI) [13]. To achieve this sensor, a $Cu_3(BTC)_2$ composite integrated with polyaniline was further coated with a thin layer of screen-printed carbon electrodes which was additionally conjugated with anti-cTnI antibodies.

Nitric oxide plays an important role in several biomedical applications. Nitric oxide, in itself, is an essential signalling molecule, found in various cardio, neural, immuno, and tumour-related physiological processes. Megan et al. reported a MOF-polymer composite compatible with human hepatocytes which can be used for the generation of nitric oxide (NO) [14]. The generation of NO is quite significant in biomedical applications as it possesses anti-inflammatory and antibacterial properties. Nevertheless, MOF-polymer composite for such applications is quite useful as they could replace the widely employed Cu-based MOFs which might induce copper leaching which is considered toxic in biomedical applications. Similarly, a Ni-based MOF integrated with polyurethane is reported to show efficient NO uptake and release [15]. Furthermore, the NO release from the composite was found to be very effective in antibacterial biomedical applications.

MOF-polymer composites have an advantage of hierarchical and controlled pore distribution which is of high importance in chemotherapeutic drug delivery. For example, a monolayer dispersion of chitosan over a solvothermally prepared cobalt-based bio-MOF is reported to be useful for treating breast cancer [16]. The corresponding drug, doxorubicin (DOX), is observed to be exhibiting useful loading and discharging characteristics within the polymer-bio-MOF composite.

MOF@polymer composites are utilized as drug bearing nano-carriers for cancer therapy and other biomedical drug delivery systems. For example, in the works of Liefeng and co-workers, a MOF (UiO) treated with methacrylic acid was coated with poly-ortho ester to form MOF@polymer composites [17]. After polymerization of MOF@polymer composite, DOX (doxorubicin) was loaded to form nano-sized drug delivery systems (DDS) useful for combination chemotherapy. In a detailed study by Barbara et al., the synergy between the hydrophilic MOF and hydrophobic polymer was studied in-situ using synchrotron micro spectroscopy [18]. The study elucidates the drug delivery mechanism in case of drug@MOF/polymer composite made using 5-Fluorouracel (5-FU) anticancer drug loaded onto HKUST-1 MOF with polyurethane. Polymer counterpart of the composites play an active role in addition to drug bearing capacity of porous MOFs. In a study of poor-water soluble drug delivery, namely indomethacin (IMC), a cyclodextrin-MOF (CD-MOF) was encapsulated with specific polymer to achieve sustained drug delivery [19]. After drug impregnation into CD-MOF, a spray drying technique was utilized to create polymer (Eudragit RS) encapsulated MOF composite to increases the solubility and dissolution of IMC drug. Similarly, Zr-based MOFs (Uio-66 and UiO-66-NH_2) loaded with ciprofloxacin drug was encapsulated with polycaprolactone using solvent casting. MOF-5 when integrated with polycaprolactone (PCL) shows promising platform to load antimicrobial drugs such cephalexin and metronidazole.

MOF-templated polymers represent modular materials characterized by a hierarchical structure with the benefits derived from the tunable structure and porosity of the MOF material combined with the flexibility and compatibility of polymers that offer unique advantages. MOFs anchored on solid surfaces (SURFMOF) will have the limitations of poor stability in physiological conditions, and the potential release of metal ions from MOF. These issues can effectively be addressed through covalent cross-linking with surface-anchored polymer gel. Furthermore, surface-anchored MOFs serve as versatile templates for polymerization via various methods including photoactivation, thermal activation, or catalytic process ensuring that the resulting polymers retain the crystalline nature of MOF material in the final polymer structure [20].

The most common synthesis route of MOF materials, solvothermal method, presents a limitation in the practical application of MOF material due to their powder form, making the processing difficult. To overcome this issue, MOF can be grown on suitably functionalized substrates by different layer-by-layer liquid epitaxy methods including spraying, spin, or automated dip coating that results in highly ordered monolithic crystalline SURMOF thin film with a smooth surface. Synthesis of SURMOF by layer-by-layer epitaxy method has added the advantage of incorporating the functional group in the pores at the desired layers by the combination of layer-by-layer epitaxial growth and post-surface modification [21–23]. Synthesis of iso reticular paddle wheel-based MOF series with tunable pore size and functionalization by liquid phase epitaxy was investigated by Liu et al. [24]. They could obtain a more symmetrical structure by this process rather than the most common energetically favourable structure. This is favoured by the carboxylic groups in each layer acting as a seed structure and resulting in a symmetrical structure with a large pore size. Further increase in pore size could be achieved by adding additional functionalities to the linker molecules. Despite the advantages of SURMOF, when biomedical applications are concerned, their cytotoxic nature is due to the presence of metal ions.

The post-synthetic modification of SURMOF, by covalent bond or coordination bond with organic linkers and metal nodes, respectively, selectively at the interior or external surfaces makes them suitable for various biomedical applications. In the process of post-synthetic polymerization of SURMOF to obtain SURGEL, the organic linker molecule with carboxylic acids and at least two orthogonally positioned functional groups is the main building block. The carboxylic groups of the main building block are coordinated with the metal node, while the orthogonally positioned functional groups are coordinated with the crosslinker molecules that interconnect the MOF structure. The interconnection of the MOF structures can happen through click chemistry [25]. The removal of metals from the polymerized SURMOF results in the formation of metal-free SURGEL. Thus, the polymerization of organic linkers of the clickable MOF material widens their biomedical applications [26]. Preassembling the organic linker molecule can control the size and shape of the SURGEL (Figure 9.1).

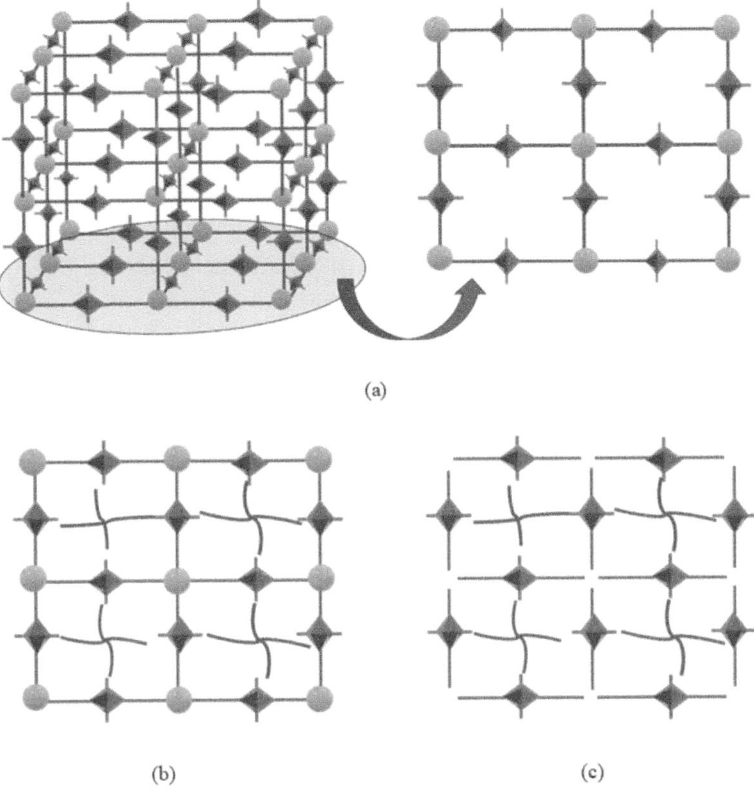

FIGURE 9.1 Schematic illustration of cross-linking of the organic linkers in MOF with crosslinker molecules and formation of polymer gel on subsequent decomposition of crosslinked MOF (a); MOF (b) organic linkers of MOF crosslinked with crosslinker molecule; and (c) MOF-derived polymer gel (SURGEL).

Polymer gel, a soft and wet elastic 3D network structure, obtained through covalent cross-linking of linear flexible polymers can also derived from hard MOF templates. Transformation of MOF material into polymer gel yields a composite material, combining the characteristics of hard MOF material and soft polymer material. This transformation occurs through the cross-linking of clickable organic linkers from MOF with the crosslinker molecules. Here, the polymerization through in situ click reactions of MOF happens between the host monomers and the guest monomer. Host monomers are carboxylate-based organic ligands with orthogonally positioned azide groups and are in the immobilized state in the framework. On the other hand, the guest monomers are mobile inside the nanopores. Multifold acetylene-tagged molecules serve as the guest monomers and are crosslinked with the host monomers. In their study, Kazuki Sada and his colleagues investigated the in situ click reactions of azide-bearing MOF material in the presence of a Cu^{2+} catalyst. The organic linkers having carboxylic acid and azide groups are coordinated with the metal node (Zn^{2+}) through the former functional group. They were crosslinked with the tetra acetylene crosslinker molecule through the orthogonally positioned azide groups. The coordination between the organic linker and the metal node gets decomposed on acid treatment with HCl resulting in MOF-templated polymer gel. Furthermore, Kazuki et al. demonstrated the click chemistry of well-defined three-dimensional porous MOF with different organic linker molecules to obtain the crosslinked MOF material. The azide containing biphenyl and triphenyl carboxylate ligands coordinated with Cu^{2+}, Zn^{2+}, and Zr^{4+} were crosslinked through click reaction to yield the crosslinked MOF. The MOF templated polymer gel with an ideal polymer network form was obtained through demetallation of crosslinked MOF by acid washing [27].

9.4 MOF-POLYMER COMPOSITE FOR ENVIRONMENTAL AND ENERGY APPLICATIONS

The polymer can be covalently attached with MOF by grafting-to grafting-from route. In the grafting-to method, the polymer is functionalized with the complementary functional group that matches the reactive group available on the grafting surface. Conversely, the grafting-from method requires functionalization of the grafting surface with the initiator group which will facilitate the growth of the polymer. The biocompatibility of the covalently attached MOF with a polymer surface renders it a promising candidate for nanoscale theranostic devices with the capability of targeted imaging and cancer treatment. Iron and gadolinium are high-spin paramagnetic metals that are used as negative and positive contrast agents, respectively. Despite the wide dynamic range of the most commonly used gadolinium-based (Gd^{3+}) contrast agent, it has limitations due to its toxicity. Rowe et al grafted multifunctional polymer chains onto gadolinium (Gd) – MOF nanoparticles. They constructed the nanoscale theranostic device by integrating a gadolinium-based MOF material as an MRI contrast agent along with other functional materials such as PFMA, GRGDS-NH_2, and MTX. These components served as a cellular-level imaging agent, targeting moiety and antineoplastic drugs, respectively. Importantly, each material preserved its functionality.

Electromagnetic pollution which is detrimental to biological and electronic equipment is mitigated by using microwave-absorbing materials with suitable dielectric and magnetic characteristics which convert the electromagnetic energy into heat energy. The microwave-absorbing material should be as thin and light as possible along with strong absorption for a wide range of frequencies. The most reported microwave-absorbing materials include metal nanoparticles, conducting polymers, metamaterials, and ceramics [28–31]. Carbonaceous materials composited with different materials including metals/metal oxides/metal sulphides, and materials with magnetic and dielectric properties exhibit enhanced microwave absorption due to the synergistic dielectric loss mechanism arising from the components of the composite. Among the various carbonaceous materials, including MWCNT, carbon fibre, activated carbon, and graphene, porous carbon derived from MOF has the inherent advantage of being lightweight due to its highly porous nature. MOF-derived pure carbon obtained by the pyrolysis of MOF material is composited with other microwave absorption material applicable for a wide frequency range. MOF-derived pure carbon composited with mono, multi-metallic components, and metal oxides exhibit better absorption than the pure carbon-based microwave absorption material. MOF-derived carbon containing the metal component, with magnetic properties, coming from the parent MOF exhibits the added advantage of tunable electromagnetic parameters by changing composition. $Cu_3(BTC)_2$-derived porous Cu/C composite was tuned to get the desirable complex permeability (μ_r) and complex permittivity (ε_r) by incorporating Co nanoparticles which are important parameters in determining the reflection loss as given in the following equations 9.1 and 9.2 [32].

$$Z_{in} = \sqrt{\mu_r/\varepsilon_r}\, \tanh\left[-j\left(\frac{2\pi f d}{c}\right)\sqrt{\mu_r/\varepsilon_r}\right] \tag{9.1}$$

$$RL = 20\log\left|\frac{Z_{in} - Z_0}{Z_{in} + Z_0}\right| \tag{9.2}$$

where Z_{in}, – input impedance, μ_r – complex permeability, ε_r – complex permittivity, d – coating thickness, f – coating thickness, c – velocity of light, Z_0 – impedance of free space.

Enhanced conduction loss due to the availability of more free electrons from metal nanoparticles and the conductive network of carbon, and strong interfacial polarization arising from the increased interfacial area between the metal nanoparticles and the carbon network show promising microwave absorption properties. The microwave absorption characteristics of pure carbon-based absorbers derived from various MOFs and various MOF-derived composites of carbon with mono, and multi-metals and metal oxides are presented in Table 9.1.

TABLE 9.1

Microwave characteristics of various MOF-derived carbon and their composites

Absorber	Reflection loss, dB (thickness in mm)	Absorption bandwidth, GHz (thickness in mm)	Ref
CNPs/GO	−66.2 (2.89)	3.6 (1.5)	[33]
Hollow GO@PC	−32.43 (3.7)	4.2 (3.5)	[34]
N-doped PC	−39.7 (4.0)	4.3 (4.0)	[35]
Fe/C	−20.3 (2.0)	7.2 (2.0)	[36]
Ni@C	−45.8 (3.5)	5.1 (3.5)	[37]
Ni/C nanosheets	−71.6 (2.2)	4.7 (2.2)	[38]
CNT/Co/C	−53.3 (2.9)	8.02 (1.6)	[39]
Fe-Co/NPC	−21.7 (1.2)	5.8 (1.2)	[40]
Co-Zn/C	−45.2 (2.5)	5.7 (2.5)	[41]
CoNi/C	−61.02 (2.0)	5.2 (2.0)	[42]
MoW-NC	−55.6 (2.8)	8.8 (2.8)	[43]
Fe_3O_4@NPC	−65.5 (3.0)	9.8 (3.0)	[44]
Fe_3O_4@NC@RGO	−72.6 (2.0)	5.5 (2.0)	[45]

9.5 CONCLUSION

Over the last decade, there has been a growing exploration of MOF-derived nanozymes, porous carbon materials, and their composites with mono, multi-metals, and metal oxides and with polymers for their application in various domains, including catalysis mimicking the natural enzymes, theranostic, imaging, and microwave absorption. The increased attention directed towards MOF-derived materials will witness significant advancements in the coming years, thanks to the commercialization of MOF. The environmentally friendly facile synthesis of MOF and MOF-derived material holds promise in addressing the challenges associated with the use of toxic solvents and energy requirements for their fabrication. This will pave the way for expanding their applications beyond current horizons.

REFERENCES

1. Shekhah, O.; Hirai, K.; Wang, H.; Uehara, H.; Kondo, M.; Diring, S.; Zacher, D.; Fischer, R. A.; Sakata, O.; Kitagawa, S.; et al. MOF-on-MOF Heteroepitaxy: Perfectly Oriented [Zn2(Ndc)2(Dabco)]n Grown on [Cu2(Ndc)2(Dabco)]n Thin Films. *Dalton Trans*, 2011, *40* (18), 4954–4958. https://doi.org/10.1039/C0DT01818J.
2. Jiang, D.; Ni, D.; Rosenkrans, Z. T.; Huang, P.; Yan, X.; Cai, W. Nanozyme: New Horizons for Responsive Biomedical Applications. *Chem Soc Rev*, 2019, *48* (14), 3683–3704. https://doi.org/10.1039/C8CS00718G.
3. Cao, F.; Zhang, Y.; Sun, Y.; Wang, Z.; Zhang, L.; Huang, Y.; Liu, C.; Liu, Z.; Ren, J.; Qu, X. Correction to Ultrasmall Nanozymes Isolated within Porous Carbonaceous Frameworks for Synergistic Cancer Therapy: Enhanced Oxidative Damage and Reduced Energy Supply. *Chem Mater*, 2022, *34* (17), 8087–8088. https://doi.org/10.1021/acs.chemmater.2c02461.
4. Li, S.; Hou, Y.; Chen, Q.; Zhang, X.; Cao, H.; Huang, Y. Promoting Active Sites in MOF-Derived Homobimetallic Hollow Nanocages as a High-Performance Multifunctional Nanozyme Catalyst for Biosensing and Organic Pollutant Degradation. *ACS Appl Mater Interfaces*, 2019, *12* (2), 2581–2590. https://doi.org/10.1021/acsami.9b20275.
5. Liu, S.; Li, Z.; Wang, C.; Tao, W.; Huang, M.; Zuo, M.; Yang, Y.; Yang, K.; Zhang, L.; Chen, S.; et al. Turning Main-Group Element Magnesium into a Highly Active Electrocatalyst for Oxygen Reduction Reaction. *Nat Commun*, 2020, *11* (1), 938. https://doi.org/10.1038/s41467-020-14565-w.
6. Wang, S.; Shang, L.; Li, L.; Yu, Y.; Chi, C.; Wang, K.; Zhang, J.; Shi, R.; Shen, H.; Waterhouse, G. I. N.; et al. Metal–Organic-Framework-Derived Mesoporous Carbon Nanospheres Containing Porphyrin-Like Metal Centers for Conformal Phototherapy. *Adv Mater*, 2016, *28* (38), 8379–8387. https://doi.org/10.1002/adma.201602197.
7. Zhou, X.; Wang, M.; Chen, J.; Xie, X.; Su, X. Peroxidase-Like Activity of Fe–N–C Single-Atom Nanozyme Based Colorimetric Detection of Galactose. *Anal Chim Acta*, 2020, *1128*, 72–79. https://doi.org/10.1016/j.aca.2020.06.027.
8. Wang, D.; Wu, H.; Phua, S. Z. F.; Yang, G.; Qi Lim, W.; Gu, L.; Qian, C.; Wang, H.; Guo, Z.; Chen, H.; et al. Self-Assembled Single-Atom Nanozyme for Enhanced Photodynamic Therapy Treatment of Tumor. *Nat Commun*, 2020, *11* (1), 357. https://doi.org/10.1038/s41467-019-14199-7.
9. Katz, M. J.; Moon, S.-Y.; Mondloch, J. E.; Beyzavi, M. H.; Stephenson, C. J.; Hupp, J. T.; Farha, O. K. Exploiting Parameter Space in MOFs: A 20-Fold Enhancement of

Phosphate-Ester Hydrolysis with UiO-66-NH2. *Chem Sci*, 2015, *6* (4), 2286–2291. https://doi.org/10.1039/C4SC03613A.

10. Karagiaridi, O.; Lalonde, M. B.; Bury, W.; Sarjeant, A. A.; Farha, O. K.; Hupp, J. T. Opening ZIF-8: A Catalytically Active Zeolitic Imidazolate Framework of Sodalite Topology with Unsubstituted Linkers. *J Am Chem Soc*, 2012, *134* (45), 18790–18796. https://doi.org/10.1021/ja308786r.

11. Sang, Y.; Cao, F.; Li, W.; Zhang, L.; You, Y.; Deng, Q.; Dong, K.; Ren, J.; Qu, X. Bioinspired Construction of a Nanozyme-Based H2O2 Homeostasis Disruptor for Intensive Chemodynamic Therapy. *J Am Chem Soc*, 2020, *142* (11), 5177–5183. https://doi.org/10.1021/jacs.9b12873.

12. Kou, D.; Ma, W.; Zhang, S.; Li, R.; Zhang, Y. BTEX Vapor Detection with a Flexible MOF and Functional Polymer by Means of a Composite Photonic Crystal. *ACS Appl Mater Interfaces*, 2020, *12* (10), 11955–11964. https://doi.org/10.1021/acs ami.9b22033.

13. Gupta, A.; Sharma, S. K.; Pachauri, V.; Ingebrandt, S.; Singh, S.; Sharma, A. L.; Deep, A. Sensitive Impedimetric Detection of Troponin I with Metal–Organic Framework Composite Electrode. *RSC Adv*, 2021, *11* (4), 2167–2174. https://doi.org/10.1039/ D0RA06665F.

14. Megan, J. N.; Brenton, R. W.; Alec, L.; Salman, R. K., Melissa, M. R. Water-Stable Metal–Organic Framework/Polymer Composites Compatible with Human Hepatocytes. *ACS Appl. Mater. Interfaces*, 2016, *8* (30), 19343–19352. https://doi.org/ 10.1021/acsami.6b05948

15. Duncan, M. J.; Wheatley, P. S.; Coghill, E. M.; Vornholt, S. M.; Warrender, S. J.; Megson, I. L.; Morris, R. E. Antibacterial Efficacy from NO-Releasing MOF–Polymer Films. *Mater Adv*, 2020, *1* (7), 2509–2519. https://doi.org/10.1039/D0MA00650E.

16. Abazari, R.; Mahjoub, A. R.; Ataei, F.; Morsali, A.; Carpenter-Warren, C. L.; Mehdizadeh, K.; Slawin, A. M. Z. Chitosan Immobilization on Bio-MOF Nanostructures: A Biocompatible PH-Responsive Nanocarrier for Doxorubicin Release on MCF-7 Cell Lines of Human Breast Cancer. *Inorg Chem*, 2018, *57* (21), 13364–13379. https://doi.org/10.1021/acs.inorgchem.8b01955.

17. Hu, L.; Xiong, C.; Wei, G.; Yu, Y.; Li, S.; Xiong, X.; Zou, J.-J.; Tian, J. Stimuli-Responsive Charge-Reversal MOF@polymer Hybrid Nanocomposites for Enhanced Co-Delivery of Chemotherapeutics towards Combination Therapy of Multidrug-Resistant Cancer. *J Colloid Interface Sci*, 2022, *608*, 1882–1893. https://doi.org/ 10.1016/j.jcis.2021.10.070.

18. Souza, B. E.; Donà, L.; Titov, K.; Bruzzese, P.; Zeng, Z.; Zhang, Y.; Babal, A. S.; Möslein, A. F.; Frogley, M. D.; Wolna, M.; et al. Elucidating the Drug Release from Metal–Organic Framework Nanocomposites via In Situ Synchrotron Microspectroscopy and Theoretical Modeling. *ACS Appl Mater Interfaces*, 2020, *12* (4), 5147–5156. https:// doi.org/10.1021/acsami.9b21321.

19. Wang, S.; Yang, X.; Lu, W.; Jiang, N.; Zhang, G.; Cheng, Z.; Liu, W. Spray Drying Encapsulation of CD-MOF Nanocrystals into Eudragit® RS Microspheres for Sustained Drug Delivery. *J Drug Deliv Sci Technol*, 2021, *64*, 102593. https://doi.org/ 10.1016/j.jddst.2021.102593.

20. Begum, S.; Hassan, Z.; Bräse, S.; Wöll, C.; Tsotsalas, M. Metal–Organic Framework-Templated Biomaterials: Recent Progress in Synthesis, Functionalization, and Applications. *Acc Chem Res*, 2019, *52* (6), 1598–1610. https://doi.org/10.1021/acs. accounts.9b00039.

21. Liu, B.; Ma, M.; Zacher, D.; Bétard, A.; Yusenko, K.; Metzler-Nolte, N.; Wöll, C.; Fischer, R. A. Chemistry of SURMOFs: Layer-Selective Installation of Functional

Groups and Post-Synthetic Covalent Modification Probed by Fluorescence Microscopy. *J Am Chem Soc*, 2011, *133* (6), 1734–1737. https://doi.org/10.1021/ja1109826.

22. Wang, S.; McGuirk, C. M.; d'Aquino, A.; Mason, J. A.; Mirkin, C. A. Metal–Organic Framework Nanoparticles. *Adv Mater*, 2018, *30* (37), 1800202. https://doi.org/10.1002/adma.201800202.

23. Cohen, S. M. Postsynthetic Methods for the Functionalization of Metal–Organic Frameworks. *Chem Rev*, 2012, *112* (2), 970–1000. https://doi.org/10.1021/cr200179u.

24. Liu, J.; Lukose, B.; Shekhah, O.; Arslan, H. K.; Weidler, P.; Gliemann, H.; Bräse, S.; Grosjean, S.; Godt, A.; Feng, X.; et al. A Novel Series of Isoreticular Metal Organic Frameworks: Realizing Metastable Structures by Liquid Phase Epitaxy. *Sci Rep*, 2012, *2* (1), 921. https://doi.org/10.1038/srep00921.

25. Islamoglu, T.; Goswami, S.; Li, Z.; Howarth, A. J.; Farha, O. K.; Hupp, J. T. Correction to Postsynthetic Tuning of Metal–Organic Frameworks for Targeted Applications. *Acc Chem Res*, 2018, *51* (1), 212. https://doi.org/10.1021/acs.accounts.7b00620.

26. Tsotsalas, M.; Liu, J.; Tettmann, B.; Grosjean, S.; Shahnas, A.; Wang, Z.; Azucena, C.; Addicoat, M.; Heine, T.; Lahann, J.; et al. Fabrication of Highly Uniform Gel Coatings by the Conversion of Surface-Anchored Metal–Organic Frameworks. *J Am Chem Soc*, 2014, *136* (1), 8–11. https://doi.org/10.1021/ja409205s.

27. Ishiwata, T.; Furukawa, Y.; Sugikawa, K.; Kokado, K.; Sada, K. Transformation of Metal–Organic Framework to Polymer Gel by Cross-Linking the Organic Ligands Preorganized in Metal–Organic Framework. *J Am Chem Soc*, 2013, *135* (14), 5427–5432. https://doi.org/10.1021/ja3125614.

28. Ding, F.; Cui, Y.; Ge, X.; Jin, Y.; He, S. Ultra-Broadband Microwave Metamaterial Absorber. *Appl Phys Lett*, 2012, *100* (10), 103506. https://doi.org/10.1063/1.3692178.

29. Li, X.; Qiao, L.; Shi, H.; Chai, G.; Wang, T.; Wang, J. Temperature Dependent Microwave Absorption Properties of SrFe12O19 in X-Band. *Front Mater*, 2022, *9*. https://doi.org/10.3389/fmats.2022.1054725.

30. Lv, H.; Yang, Z.; Pan, H.; Wu, R. Electromagnetic Absorption Materials: Current Progress and New Frontiers. *Prog Mater Sci*, 2022, *127*, 100946. https://doi.org/10.1016/j.pmatsci.2022.100946.

31. Li, X.; Wang, L.; You, W.; Xing, L.; Yu, X.; Li, Y.; Che, R. Morphology-Controlled Synthesis and Excellent Microwave Absorption Performance of ZnCo2O4 Nanostructures via a Self-Assembly Process of Flake Units. *Nanoscale*, 2019, *11* (6), 2694–2702. https://doi.org/10.1039/C8NR08601J.

32. Liu, W.; Liu, L.; Yang, Z.; Xu, J.; Hou, Y.; Ji, G. A Versatile Route toward the Electromagnetic Functionalization of Metal–Organic Framework-Derived Three-Dimensional Nanoporous Carbon Composites. *ACS Appl Mater Interfaces*, 2018, *10* (10), 8965–8975. https://doi.org/10.1021/acsami.8b00320.

33. Zhao, H.; Han, X.; Li, Z.; Liu, D.; Wang, Y.; Wang, Y.; Zhou, W.; Du, Y. Reduced Graphene Oxide Decorated with Carbon Nanopolyhedrons as an Efficient and Lightweight Microwave Absorber. *J Colloid Interface Sci*, 2018, *528*, 174–183. https://doi.org/10.1016/j.jcis.2018.05.046.

34. Xu, H.; Yin, X.; Zhu, M.; Li, M.; Zhang, H.; Wei, H.; Zhang, L.; Cheng, L. Constructing Hollow Graphene Nano-Spheres Confined in Porous Amorphous Carbon Particles for Achieving Full X Band Microwave Absorption. *Carbon N Y*, 2019, *142*, 346–353. https://doi.org/10.1016/j.carbon.2018.10.056.

35. Wu, Q.; Jin, H.; Chen, W.; Huo, S.; Chen, X.; Su, X.; Wang, H.; Wang, J. Graphitized Nitrogen-Doped Porous Carbon Composites Derived from ZIF-8 as Efficient Microwave Absorption Materials. *Mater Res Express*, 2018, *5* (6), 065602. https://doi.org/10.1088/2053-1591/aac67e.

36. Qiang, R.; Du, Y.; Zhao, H.; Wang, Y.; Tian, C.; Li, Z.; Han, X.; Xu, P. Metal Organic Framework-Derived Fe/C Nanocubes toward Efficient Microwave Absorption. *J Mater Chem A Mater*, 2015, *3* (25), 13426–13434. https://doi.org/10.1039/C5TA01457C.

37. Yan, J.; Huang, Y.; Yan, Y.; Ding, L.; Liu, P. High-Performance Electromagnetic Wave Absorbers Based on Two Kinds of Nickel-Based MOF-Derived Ni@C Microspheres. *ACS Appl Mater Interfaces*, 2019, *11* (43), 40781–40792. https://doi.org/10.1021/acs ami.9b12850.

38. Wen, B.; Yang, H.; Lin, Y.; Wang, L.; Ma, L.; Qiu, Y. In Situ Anchoring Carbon Nanotubes on the Ni/C Nanosheets with Controllable Thickness for Boosting the Electromagnetic Waves Absorption. *Compos Part A Appl Sci Manuf*, 2020, *138*, 106044. https://doi.org/10.1016/j.compositesa.2020.106044.

39. Yang, M.; Yuan, Y.; Li, Y.; Sun, X.; Wang, S.; Liang, L.; Ning, Y.; Li, J.; Yin, W.; Che, R.; et al. Dramatically Enhanced Electromagnetic Wave Absorption of Hierarchical CNT/Co/C Fiber Derived from Cotton and Metal-Organic-Framework. *Carbon N Y*, 2020, *161*, 517–527. https://doi.org/10.1016/j.carbon.2020.01.073.

40. Zhang, X.; Ji, G.; Liu, W.; Quan, B.; Liang, X.; Shang, C.; Cheng, Y.; Du, Y. Thermal Conversion of an Fe3O4@metal–Organic Framework: A New Method for an Efficient Fe–Co/Nanoporous Carbon Microwave Absorbing Material. *Nanoscale*, 2015, *7* (30), 12932–12942. https://doi.org/10.1039/C5NR03176A.

41. Pan, J.; Xia, W.; Sun, X.; Wang, T.; Li, J.; Sheng, L.; He, J. Improvement of Interfacial Polarization and Impedance Matching for Two-Dimensional Leaf-like Bimetallic (Co, Zn) Doped Porous Carbon Nanocomposites with Broadband Microwave Absorption. *Appl Surf Sci*, 2020, *512*, 144894. https://doi.org/10.1016/j.apsusc.2019.144894.

42. Wang, Y.-L.; Yang, S.-H.; Wang, H.-Y.; Wang, G.-S.; Sun, X.-B.; Yin, P.-G. Hollow Porous CoNi/C Composite Nanomaterials Derived from MOFs for Efficient and Lightweight Electromagnetic Wave Absorber. *Carbon N Y*, 2020, *167*, 485–494. https://doi.org/10.1016/j.carbon.2020.06.014.

43. Liu, P.; Gao, S.; Huang, W.; Ren, J.; Yu, D.; He, W. Hybrid Zeolite Imidazolate Framework Derived N-Implanted Carbon Polyhedrons with Tunable Heterogeneous Interfaces for Strong Wideband Microwave Attenuation. *Carbon N Y*, 2020, *159*, 83–93. https://doi.org/10.1016/j.carbon.2019.12.021.

44. Xiang, Z.; Song, Y.; Xiong, J.; Pan, Z.; Wang, X.; Liu, L.; Liu, R.; Yang, H.; Lu, W. Enhanced Electromagnetic Wave Absorption of Nanoporous Fe3O4 @ Carbon Composites Derived from Metal-Organic Frameworks. *Carbon N Y*, 2019, *142*, 20–31. https://doi.org/10.1016/j.carbon.2018.10.014.

45. Xiang, Z.; Xiong, J.; Deng, B.; Cui, E.; Yu, L.; Zeng, Q.; Pei, K.; Che, R.; Lu, W. Rational Design of 2D Hierarchically Laminated Fe3O4@nanoporous Carbon@ rGO Nanocomposites with Strong Magnetic Coupling for Excellent Electromagnetic Absorption Applications. *J Mater Chem C Mater*, 2020, *8* (6), 2123–2134. https://doi. org/10.1039/C9TC06526A.

Index